The Way of Data

of

Data

From Technology
to Applications

The Way of Data

From Technology to Applications

Hequan Wu

China Information and
Communication Technology Group, China

World Scientific

NEW JERSEY · LONDON · SINGAPORE · BEIJING · SHANGHAI · HONG KONG · TAIPEI · CHENNAI · TOKYO

Published by

World Scientific Publishing Co. Pte. Ltd.

5 Toh Tuck Link, Singapore 596224

USA office: 27 Warren Street, Suite 401-402, Hackensack, NJ 07601

UK office: 57 Shelton Street, Covent Garden, London WC2H 9HE

Library of Congress Cataloging-in-Publication Data
Names: Wu, Hequan, author.
Title: The way of data : from technology to applications / Hequan Wu,
    China Information and Communication Technology Group, China.
Description: New Jersey : World Scientific, [2023] | Includes bibliographical references and index.
Identifiers: LCCN 2023000216 | ISBN 9789811250071 (hardcover) |
    ISBN 9789811250088 (ebook) | ISBN 9789811250095 (ebook other)
Subjects: LCSH: Big data--Industrial applications. | Big data--Economic aspects. |
    Big data--Social aspects.
Classification: LCC QA76.9.B45 W823 2023 | DDC 005.7--dc23/eng/20230127
LC record available at https://lccn.loc.gov/2023000216

British Library Cataloguing-in-Publication Data
A catalogue record for this book is available from the British Library.

B&R Book Program

For any available supplementary material, please visit
https://www.worldscientific.com/worldscibooks/10.1142/12664#t=suppl

Translator: ZHAO Chao @lan-bridge
Desk Editors: Logeshwaran Arumugam/Steven Patt

Typeset by Stallion Press
Email: enquiries@stallionpress.com

Preface

The third industrial revolution, characterized by information technology, has been more than 30 years since the beginning of Internet commerce. In these years, digitalization has constantly penetrated all aspects of social life. The transformation of digitalization is going deep into all kinds of industries, from the application of digital technology to the mining of data value. The role of data as a new factor of production is increasingly prominent. Big data technology and application level have become an important indicator of the competitiveness of governments and enterprises in today's society.

The chain of data value is very long, covering the whole process of data perception, collection, transmission, storage, calculation, transaction, processing, supervision, analysis, decision-making, etc., involving the Internet of things, 4G/5G, cloud computing, computing power network, artificial intelligence, information security, and other technical fields. Big data mining from technology to application is a systematic project. Big data technology has developed rapidly with the deepening of application, learning and applying big data require clear thinking, and it is particularly necessary to explore the way of data. Understand the way of data, learn to discover the rules of things from data, excavate the connotation and value, improve the pertinence and accuracy of social governance, improve production efficiency and product quality, and improve service level and efficiency.

Academician Zheng Weimin of Tsinghua University recommended the topic of Way of Data to China Science and Technology Press, guided the design of the outline of the book, and was responsible for quality control. Professor Liu Peng of Tsinghua University was responsible for chapter design and unified drafting. Professor Zhang Yan wrote Chapters 1, 2, 5, 7, and 8. Professor Liang Nan wrote Chapters 4 and 9. Professor Wu Zhenghao wrote Chapters 3 and 6. Mr. Zheng Hongwei and Ms. Li Jie of the China Science and Technology Press, as planning editors and responsible editors respectively, carefully organized the editing. The Chinese version of this book, *The Way of Data: From Technology to Applications*, was published by the China Science and Technology Press in 2019. This book is positioned as a popular science book, introducing the basic knowledge of big data, focusing on the examples and application prospects of big data mining from government, industry, people's livelihood, and other aspects. This book explains the way of data in popular language so that the public readers can understand the role of data and the method of applying data. China Science and Technology Press also introduced this book to the World Science and Technology Press. Mr. Zhao Chao was responsible for the translation of this book, and the translation of the book was assisted by Lan-Bridge Translation Company. Steven Shi Hongbing, Steven Patt, and Logeshwaran Arumugam of World Scientific Publishing worked carefully as editors of the English version of this book to make the English version published. This book lists the references and cites the views and figures of these references. The authors of this book would like to thank the authors of the cited references. In addition, the compilation and drawing of this book have been helped by some students. Without their support, it is impossible to complete the publication of this book. I sincerely thank the experts, teachers, students, and editorial department who supported and helped the publication of this book.

About the Authors

Wu Hequan is a member of the Chinese Academy of Engineering (CAE) and the vice president of CAE from 2002 to 2010. He once served as the vice president and chief engineer of the China Academy of Telecommunications Technology (CATT). He has been engaged in the research and development of digital communication and optical communication systems. He has been the director of Expert Committee of China Next Generation Internet (CNGI) demonstration project in 2003–2016, the director of the National IPv6 Scale Deployment Expert Committee, and the chief engineer of China's national new generation mobile communication network major S&T projects in 2016–2020. From 2013 to 2019, he served as the president of the Internet Society of China (ISC). He is now the director of the Advisory Committee of ISC. He is also the director of China's Standardization Expert Committee since 2014.

Zheng Weimin is a CAS member. He is currently a professor and a doctoral supervisor at the Department of Computer Science and Technology, Tsinghua University. He was the President of the China Computer Federation. Prof. Zheng has long been engaged in the research of high-performance computer architecture as well as parallel algorithms and systems. He led the establishment and application of the cluster architecture of high-performance computers in China and participated in the development of the extremely large-scale weather forecast application based on the domestic Sunway TaihuLight, which won the ACM Gordon Bell Award in 2016. He has served as the director of the 863 High-performance Computer Evaluation Center. His contributions to some scientific problems and engineering techniques such as the scalability, reliability, and cost-efficiency of storage systems are highly praised by both domestic and international peers, and the network storage system, disaster-tolerant system, and self-maintenance system developed by his research team are playing important roles in multiple grand projects.

He received the Beijing Excellent Teacher Award, and the title of Beijing Famous Teacher, Special Allowance of the State Council, the State Science and Technology Progress Award (one 1st and two 2nd prizes), the State Technological Invention Award (2nd prize), He Liang He Li Science and Technology Progress Award, and the first China Storage Lifetime Achievement Award. Prof. Zheng and his collaborators published more than 500 papers and more than 10 books. The course on Computer Architecture given by Prof. Zheng was selected as a quality course at Tsinghua University and was selected as a national quality course in 2008. He is now the Editor in Chief of the journal *Big Data*.

Contents

Chapter 1

Getting to Know Big Data

If we want to select today's hottest 10 technologies, big data is undoubtedly a strong "seeded player". Big data has penetrated into all walks of life, where the emerging technologies such as driverless vehicles, intelligent transportation, and intelligent medical care are all relying on "big data" as an information asset in some sense, and more and more fields will be in this "data storm". Although everyone is talking about big data, most people know very little about its "ins and outs".

What Is Big Data

Big data is not only a technology but also a business model, and is building a new ecology that is slowly changing our lives. The concept of big data has been widely discussed, and many people think that the connotation of "big data" actually lies in the fact that "the dataset is big". For this reason, it is imperative to clarify the concept and characteristics of big data.

The definition of big data

Big data, as described by Gartner, the world's leading IT research and advisory firm, is a massive, high-growth, and diverse information asset that requires new processing models to provide greater decision-making, insight discovery, and process optimization.

Big data is defined by the McKinsey Global Institute (MGI) in *Big Data: The Next Frontier for Innovation, Competition, and Productivity* as a dataset whose size exceeds the capacity of typical database software that could capture, store, manage, or analyze. Since then, big data has gained widespread global attention from an economic perspective.

Dr. Tu Zipei, an information scholar in China, equates big data with traditional small data (derived from measurement) + modern big records (derived from records). Among them, records come from pictures, audio, video, etc., and with the gradual increase of records, big data is getting "bigger".

On August 31, 2015, the State Council of the People's Republic of China issued the *Action Plan for Promoting the Development of Big Data*, which pointed out that "big data is a data collection characterized by large capacity, multiple types, fast access, and high application value. It is rapidly developing into a new generation of information technology and service format that collects, stores, and correlates data of a huge amount, scattered sources, and various formats, whereby we can discover new knowledge, create new value, and enhance new capabilities".

The *White Paper on Big Data (2016)* prepared by the China Academy of Information and Communications Technology proposes that "big data is a hybrid of new resources, new technologies, and new ideas. From the resource perspective, big data is a new resource that represents a new view of resources; from the technical perspective, big data represents a new generation of data management and analysis technology; from the conceptual perspective, big data opens up a new perspective of thinking".

So far, it is still difficult for the academic community to give a precise definition of the technical concept of "big data". Since there is no unified definition of big data, we may understand it as a resource, a tool, and a concept of thinking and understanding the world.

Talking with big data

In daily life, the onslaught of big data can be overwhelming. As a technology, tool, and method, big data is increasingly influential and impactful on modern social life. In some fields, it is even revolutionary and disruptive. Throughout the history of human technological development, it seems that there is no technological revolution like "big data" that only took a few years that have developed from conception to the current outbreak states.

Talking with data. In the era of big data, "everything is data", and this big data era is focusing on "quantifying everything". Human beings live in a massive, dynamic, and diverse data world where data are used by everyone, anywhere and anytime. Data are as common as sunlight, air, and water, and as important as a magnifying glass, a telescope, and a microscope. In the past, we work by experience, but now, we must learn to speak with data.

Let the data speak for itself. In the ocean of all-embracing data, we will find that there are many glittering treasures hidden in the data, and through picking multiple things that previously seemed unrelated, we can uncover the hidden interrelationships between things, predicting the future while helping people to recognize things and take control of the situation. This is the key potential and value of big data.

In the process of exploring the value of data, we have gone beyond the search for cause-and-effect relationships and have expanded the scope to look at all kinds of things that are universally connected. In this process, the correlation becomes the focus of exploration, and the *what* is somewhat more important than the *why*. As the famous big data expert Viktor Mayer-Schönberger said, "correlation, not causation" has become a distinctive feature of the big data era.

For correlation, there are many examples in daily life that can help understand. For example, in a business scenario, analysis shows that most customers buy steak along with some pepper because the two are a common combination on the dinner table. Therefore, shopping malls will greatly increase sales revenue by placing shelves selling the two items together and offering coupons for matching sales.

For steak and pepper, it is easier to associate them together. However, beer and diapers, which are two seemingly unrelated items, are often found in the shopping baskets of Walmart customers in the United States at the same time.

After investigation and analysis, we learn that in American families with babies, the mother usually takes care of the child at home while the father goes shopping for diapers. During the purchase process, the father often buys a few bottles of beer for himself while buying diapers. Over time, there is always a scene of beer and diapers in the same basket.

So, Walmart, which discovers this pattern, places beer and diapers on adjacent shelves to boost sales.

In addition to its ability to be used in business scenarios, big data can also predict flu epidemics in advance. Typically, before the flu hits, there is a significant increase in the number of people searching online for relevant symptoms of illness. In 2008, Google launched Google Flu Trends (GFT), which estimated the current global epidemic in near real time based on aggregated Google search data. In 2009, Google accurately predicted the spread of H_1N_1 in the United States through epidemic prediction. This is the huge power of correlation.

"Talking with data" and "letting data speak" have become brand-new ways for human beings to perceive the world. The world is material, and material is data. Data are redefining the material origin of the world and giving "seeking truth from facts", a new connotation of the times. We must be good at speaking with data, making decisions with data, managing with data, and living with data.

As an emerging factor of production, corporate capital, and social wealth, big data is inexhaustible and can be reused and recycled. It can be said that big data is a rich mine of information and knowledge, which contains unlimited business opportunities and huge benefits. Also, there will always be unexpected gains as long as you conduct in-depth analysis and mining. The successful practice and brilliant performance of leading companies such as Google, Amazon, Facebook, Alibaba, Tencent, and JD.com are the most vivid and powerful examples.

Whoever gets the data wins the world. In addition to business opportunities and benefits, big data is also "the oil of the future" and will become the driving force for social innovation and development. Big data is promoting the transformation and reform of scientific research paradigms, industrial development models, social organization forms, and national governance methods. "Data can govern and strengthen the country"; big data has great potential in China. China is a country with a large population, a large manufacturing industry, and a large number of Internet users. These are the most active data generators. According to International Data Corporation (IDC), the scale of the digital universe is expected to reach 40 ZB by 2020, and China will generate 21% of the world's data. The good news is that China has made strategic deployment of big data, and formulated development plans and action plans. We can race on the same starting line as developed countries and may achieve overtaking on the curve.

To borrow the warning of Viktor Mayer-Schönberger and Kenneth Cukier: In the era of big data, if we refuse to embrace it, individuals may lose their lives, and countries may lose their future, as well as the future of a generation.

China's strategies for developing big data

As big data has increasingly become the means of production and value assets, accelerating the deployment of big data and deepening the application of big data have become major national events to seize the opportunity of the "data revolution". As General Secretary Xi Jinping emphasized, "the opportunity is fleeting. If we could take the chance, it will be an opportunity; If not, it will be a challenge". In order to take this revolutionary chance, China has long carried out policy planning and elevated the development of big data to a national strategy.

In July 2012, the State Council issued the *Twelfth Five-Year Plan for the Development of National Strategic Emerging Industries*, which clearly proposed to support the R&D and industrialization of massive data storage and processing technologies. In January 2013, five ministries including the Ministry of Industry and Information Technology, the National Development and Reform Commission, and the Ministry of Land and Resources jointly issued the *Guiding Opinions on the Construction and Layout of Data Centers*, which pointed out the direction for the future development of China's data centers and provided safeguards for the construction and layout of data centers.

In August 2015, the State Council issued the *Outline of Action for Promoting the Development of Big Data*, clearly proposing to build China into a data-powerful country. In March 2016, the *Outline of the Thirteenth Five-Year Plan for National Economic and Social Development of the People's Republic of China* clearly implemented the national big data strategy. In 2017, the Ministry of Industry and Information Technology issued the *Big Data Industry Development Plan (2016–2020)* and ultimately positioned the development goal as follows: "By 2020, a big data industry system with advanced technology, prosperous applications, and strong guarantees should take shape," so as to build the top-level architectural design framework of China's big data industry.

At present, in terms of technical advantages, China already has the technical and industrial foundation for the development of big data.

Among the top 10 Internet companies in the world, China occupies four seats, especially in vertical application fields, such as smart logistics and mobile payment. Even the United States, which is relatively advanced in the core technology of big data, is inferior to China in these fields.

At the same time, due to the late start, the limitations of China's big data development cannot be ignored. Compared with developed countries, China's core technologies such as new computing platforms, distributed computing architecture, big data processing, analysis, and presentation still fall behind the foreign countries.

Foreign strategies for developing big data

Since 2008, when *Nature* magazine brought the term "big data" into the public eye, it only took 10 years for it to play an important role in all walks of life. In the past 10 years, most of the countries in the world are in the initial stage of big data development. Even developed countries and regions such as the United States, Japan, and the European Union are at almost the same starting point in the development of the emerging technology of big data.

The United States was an early responder to the wave of big data. In 2009, in order to accelerate the openness and sharing of the public sector, the United States made great efforts to build the Data.gov portal, which integrates data from 50 departments including finance, medical care, science and education, transportation, and energy. Through the construction and improvement of the OGPL platform, systems such as data exchange, user communication, and data resource management are formulated to further improve the platform functions.

In March 2012, the US government announced the implementation of the *Big Data Research and Development Initiative*, integrating six federal government departments including the Department of Energy, the National Science Foundation, the National Institutes of Health, and the Department of Defense to form the Big Data Senior Steering Group. The United States government invested $200 million US dollars to fund big data research and development programs to strengthen the big data application capabilities of enterprises, governments, and academia, and also to improve the ability to collect and obtain valuable information from massive data, and to improve the level of research and development and security in scientific research, education, and national security in the era of big data.

In addition, under the guidance and support of the United States government, companies such as Google, Oracle, IBM, Facebook, and other enterprises attach great importance to the research and development of big data technology, and have tried to apply a variety of methods and measures to increase research and development and promotion efforts in data collection, cleaning, mining, analysis, and visualization. Intensified R&D and promotion efforts have further accelerated and boosted the process of big data industrialization and marketization.

The United Kingdom has always regarded the big data industry as a new economic growth point, hoping to stimulate its own economic development. In 2013, the British government released the *Strategic Plan for the Development of British Data Capability*. In order to further promote the development of big data technology, the government invested 189 million pounds and then put out 73 million pounds into the development of big data technology. It is mainly used to carry out the application of big data technology in 55 government data analysis programs. The programs rely on higher education institutions to invest in the establishment of big data research centers, and to lead famous universities including Oxford University and the University of London to open big data-focused majors.

Similar to the Data.gov platform in the United States, the British government has established the data.gov.uk website, known as the British Data Bank, which publishes the government's public government information, and provides an official outlet for the public to easily retrieve, access, and verify government data and information through a public platform. It is worth noting that the world's first non-profit Open Data Institute (ODI), which uses Internet technology to aggregate data from around the world, was also established in Britain.

Japan regards emerging industry clusters derived from big data and cloud computing as an important starting point for boosting economic growth and optimizing national governance. In June 2013, the Japanese government officially announced its new IT strategy, the *Declaration to be the World's Most Advanced IT Nation*. The national IT strategy, centered on open big data, plans to build Japan into a society with the highest level of information technology in the world. The Japanese government states that in principle, by 2020, all government information systems will be cloud-based, which will reduce operating costs by 30%.

When it comes to the leading edge of the big data industry, different countries have different advantages and priorities. Pan Wen, director of the Software Institute of the CCID Research Institute of the Ministry of Industry and Information Technology, pointed out that "at present, the United States, the United Kingdom, France, Australia, and other countries are in a leading position in the core technology of big data". At the same time, in terms of data protection, Europe maintains a leading position, Japan successfully applies big data to medical and transportation fields, and Singapore is unique in e-government.

The Source of Big Data

In 1965, Gordon Moore, the founder of Intel, introduced "Moore's Law", which states that the number of transistors that can fit on an integrated circuit increases by a factor of one approximately every 18 months, and performance increases by a factor of one when the price remains the same. In 1998, Turing Award winner Jim Gray proposed the "New Moore's Law", which states that the total amount of data in human history will increase by a factor of 1 every 18 months.

As can be seen in Figure 1.1, the total amount of global data was 30 EB (1 EB = 1,024 PB) in 2004, 50 EB in 2005, 161 EB in 2006, and 7,900 EB in 2015. According to forecasts, it will reach 35,000 EB in 2020.

Listed in the following is a set of Internet data in 2016 to show how big "big data" really is.

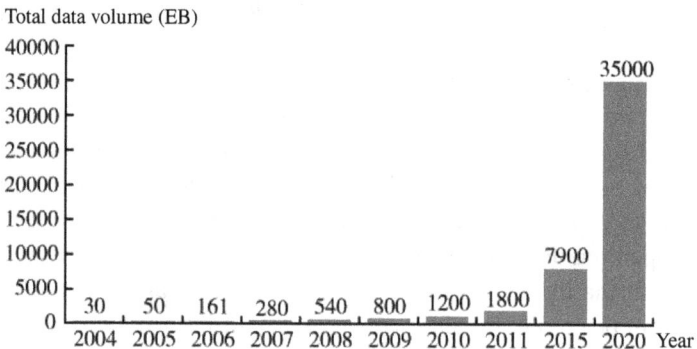

Figure 1.1   The Total Amount of Global Data

(1) All the content generated by the Internet every day can fill 640 million DVDs.
(2) Google needs to process 24 PB of data every day.
(3) On Facebook, Internet users spend 700 billion minutes every day, and they send and receive up to 1.3 EB of data.
(4) 50 million messages are posted on Twitter every day. Assuming one message takes 10 seconds to view, it will take around 16 years day and night for one person to browse all the messages.

There are three main reasons for generating such a huge amount of data.

One is the massive data generated by a large number of people. About 3 billion people in the world have access to the Internet. In the era of Web 2.0, everyone is not only a recipient of information but also a producer of information. They are a huge source of data since almost everyone is using smart terminals to take pictures, shoot videos, send microblogs, send WeChat, etc.

The second is the massive amount of data generated by a large number of sensors. At present, there are 3–5 billion sensors worldwide, and it is expected to reach 10 trillion by 2020. These sensors generate data 24 hours a day, leading to an explosion of information.

Third, scientific research and various industries are increasingly relying on big data to carry out their work. For example, the Large Hadron Collider at the European Institute of Particle Physics needs to process 100 PB of data per year, and the annual increase is 27 PB; the petroleum sector uses seismic exploration methods to detect geological structures and find oil, which requires a large number of sensors to collect seismic waveform data; to ensure the safety of high-speed rail operation, a large number of sensors need to be deployed around the rails to sense foreign objects, landslides, flooding, deformation, earthquakes, and other anomalies.

That is to say, with the further expansion of human activities, the scale of data will expand dramatically. In various fields, including finance, automobiles, retail, catering, telecommunications, energy, government affairs, medical care, sports, and entertainment, the rapid accumulation of data and the increasingly complex data types have surpassed the capabilities of traditional data management systems and processing models, so the concept of "big data" came into being.

From another perspective, big data is a wide range of structured or unstructured data through various data collectors, databases, open-source data releases, GPS information, network traces (such as shopping and search history), sensors collection, and user saving and uploading. We can classify the sources of big data from three aspects: the subject that generates the data, the sectors of the data, and the form of data storage.

By the subject that generates the data

It can be divided into three aspects according to the subjects that generate data: (1) data generated by a small number of enterprise applications, such as data in relational databases and data in data warehouses; (2) data generated by a large number of people, such as Twitter, Weibo, communication software, mobile communication data, e-commerce online transaction log data, and comment data related to enterprise applications; (3) data generated by huge amounts of machines, such as application server logs, various types of sensor data, image and video surveillance data, QR code, and barcode scanning data.

By the sectors of the data

According to the data source, the sector can be divided into five aspects: (1) Internet companies represented by the three major Internet companies (BAT); (2) telecommunications, finance, insurance, electric power, and petrochemical industries; (3) public safety, medical care, and transportation fields; (4) meteorology, geography, government affairs, and other fields; (5) manufacturing and other traditional industries.

The first is the Internet companies represented by BAT. The total amount of Baidu's data exceeds the level of 1,000 PB, covering Chinese web pages, Baidu promotion, Baidu logs, UGC, and many other parts, and sits on a huge search data with more than 70% of the search market share. Alibaba Company stores more than 100 PB of data, and owns more than 90% of the e-commerce data, including click data, user browsing data, transaction data, and shopping data. The total amount of data stored by Tencent is still more than one hundred petabytes after being compressed, and the amount of data has increased by 10% every month, including a large amount of text, audio, video, and relational data accumulated in social, gaming, and other fields.

The second is the telecommunications, finance, insurance, electric power, and petrochemical industries. Telecom industry data includes users' online records, calls, information, and geographic location data. The amount of data owned by operators is nearly one hundred petabytes, and the annual user data growth exceeds 10%. Finance and insurance include account opening information data, bank network data, online transaction data, and self-operation data. The financial system generates more than tens of petabytes of data every year, and the data volume of the insurance system also exceeds the petabyte level. In the power and petrochemical industries, the total amount of data collected by the State Grid alone reaches tens of petabytes, and the petrochemical industry generates and stores nearly one hundred petabytes of data every year.

The third is the field of public safety, medical care, and transportation. In a large and medium-sized city, the number of traffic checkpoint records in a month can reach 300 million. The data that the entire medical and health industry stores in a year can reach the level of hundreds of petabytes. The data generated by a round-trip flight reach the terabyte level. The video and text data generated and stored by water and land transport also reach tens of beat bytes every year.

The fourth is the field of meteorology, geography, and government affairs. The data stored by the China Meteorological Administration is nearly 10 PB, with an annual increase of hundreds of terabytes. The maps and geographic location information are dozens of petabytes per year. The government affairs data cover tourism, education, transportation, medical care, and other categories, and are mostly structured data.

The fifth is the field of manufacturing and other traditional industries. The types of big data in the manufacturing industry are mainly product design data, business data, and production monitoring data in the production process of enterprises. Among them, product design data is mainly file-based, and unstructured, with high sharing requirements and long storage time. The business data of enterprise production link is mainly database structured data. The production monitoring data has a very large data volume. In other traditional industries, although the data volume in offline commercial sales, agriculture, forestry, animal husbandry, and fishery, offline catering, food, scientific research, logistics, and transportation industries has increased dramatically, the amount of data is still in the accumulation period, and the overall volume is not large, mostly in the range from tens of terabytes to hundreds of terabytes.

By forms of data storage

Big data is not only reflected in the large amount of data but also in the variety of data types. Among such massive data, only about 20% is structured data, and the remaining 80% is unstructured data, widely existing in social networks, the Internet of Things (IoT), e-commerce, and other fields.

Structured data is simply data stored in databases, such as enterprise ERP, financial systems, medical HIS database, education card, government administrative approval, and other core databases.

Unstructured data includes all formats of office documents, text, images, XML, HTML, various reports, images, audio, video information, and other data.

The Six Characteristics of Big Data

In 2001, Douglas Laney, an analyst of Metta Group, first proposed "volume", "velocity", and "variety" as three key characteristics of big data in *3D Data Management: Controlling Data Volume, Velocity, and Variety*. On this basis, IBM proposed the "4V" characteristics of big data that are now recognized by the industry: volume (large volume), variety (wide variety), velocity (fast speed), and value (high value).

Undoubtedly, these four points are indeed very important characteristics of big data. However, as time goes by and as big data evolves, more big data characteristics have emerged, such as variability and complexity.

Large volume

Big is the main characteristic of big data. From the beginning of written records to the beginning of the 21st century, the total amount of data generated by humans cumulatively is only equivalent to the amount of data created by the world in just one or two days now, "1 day = 2,000 years". According to IDC's report, the data stored worldwide in 2013 are expected to reach 1.2 ZB (1 ZB = 1,024 EB). If it is stored on a CD-ROM and divided into five piles, each pile can extend to the moon. From 2013 to 2020, the scale of human data will expand 50 times, and the amount of data generated each year will grow to 44 ZB, which is equivalent to millions of times the data volume of the US National Library, and doubles every 18 months.

Wide variety

Compared with traditional data, big data has a wide range of data sources, multiple dimensions, and various types. While various machines and instruments automatically generate data, people's own life behaviors are also constantly creating data. Not only business data within enterprise organizations but also a huge amount of relevant external data. In addition to structured data such as numbers and symbols, there is also a large amount of unstructured data including web logs, audio, video, pictures, and geographic location information, accounting for more than 90% of the total data.

Fast speed

With the development of modern sensing, the Internet, and computer technologies, the speed of data generation, storage, analysis, and processing is far beyond what people can imagine. This is a significant feature that distinguishes big data from traditional data or small data. For example, the ion collider at the European Center for Nuclear Research (CERN) generates up to 40 TB of data per second of operation; 1 Boeing jet engine generates 10 TB of operational data every 30 minutes; Facebook has 1.8 billion photos uploaded or spread every day. The 3 billion base pairs of human genes that have been deciphered in the past 10 years can now be completed in just 15 minutes. In 2016, among the top 500 supercomputers in the world announced by the International Supercomputing Conference (ISC) in Frankfurt, Germany, "Sunway TaihuLight" developed by China's National Supercomputing Wuxi Center won the first place, with a peak performance of 125 PFLOPS, and its computing power of 1 minute is equivalent to 7 billion people around the world using calculators to calculate continuously for 32 years at the same time.

High value

As a basic strategic resource, in addition to basic data applications, big data can be applied in various industries such as government affairs, finance, transportation, and environmental protection through in-depth data analysis and mining, and further play the role of value assignment and empowerment. In short, big data has the characteristics of high value.

While helping to optimize resource allocation, it can further assist decision-making, improve decision-making ability, and truly change people's way of life.

Variability

Big data includes structured, semi-structured, and unstructured data. Whether it is structured data stored in a database to achieve logical expression through a two-dimensional table structure, unstructured data that cannot be expressed using a two-dimensional logical table in a database, or semi-structured data with a variable number of fields that can be extended as needed, all make the data structure describing phenomena, relationships or logic present variable forms and types. As time passes or when triggered by certain events, the unstable data flow presents the fluctuating characteristics, and the irregularity and variability of the data will become increasingly prominent.

Complexity

Affected by many factors such as data structure, type, size, and change, big data presents a certain complexity, and data extraction, data loading, data conversion, data association, and other links are becoming increasingly complex. In particular, with the explosive growth of data, it has become more and more challenging to quickly and effectively extract big data information and realize the transformation from "data" to "knowledge".

From the Internet of Things to Artificial Intelligence

The "big data pyramid" composed of the IoT, cloud computing, big data, and artificial intelligence is quietly changing people's lives. The four are closely connected and inseparable, forming a tightly interlocked whole. The IoT can widely perceive and collect all kinds of data, and play a role in data acquisition. Cloud computing can provide the storage and processing capacity of big data and play the role of data bearing. Big data technology can manage and mine big data, extract information from data, and play a key role in data analysis. Artificial intelligence can learn data, turn data into knowledge, and play the role of data understanding. These four

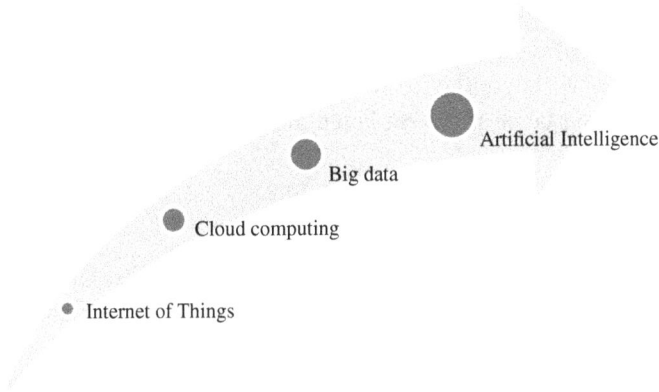

Figure 1.2   The IoT, Cloud Computing, Big Data, and Artificial Intelligence are Progressive and Interdependent

form a progressive and interdependent relationship (Figure 1.2). As academician Wu Hequan said, "We are now entering an era of big data, intelligence, IoT, mobile Internet, and cloud computing".

Internet of Things

Every morning when we wake up, we will habitually click on the weather app to check the weather condition of the day. When we are ready to go out, we will open the taxi and/or map software to call a taxi or navigate to the destination with one click. During exercising, we use various sports bracelets and cell phone apps to help record the amount of exercise and movement trajectory. With the opening of new applications such as wearables, the Internet of Vehicles, and smart meter reading, the era of the Internet of Everything is accelerating, and the connection between things has entered the fast lane of growth. According to the Global System for Mobile Communications Association (GSMA), there will be 27 billion connected devices by 2020.

The IoT can be simply understood as a network of things connected to things. Sensors, positioning systems, and sensing devices deployed in various places, in accordance with the agreed protocols, communicate with each other through the Internet for identification, positioning, tracking, and management. This will realize information exchange and communication, and eventually achieve "the connection of everything".

At present, the key technologies of IoT include sensor technology, radio frequency identification (RFID) tag, and embedded system technology. The IoT industry is composed of five layers: application layer, support layer, perception layer, platform layer, and transmission layer.

Cloud computing

The growth, use, and delivery model of Internet-based related services in cloud computing usually involves the provision of dynamically scalable and often virtualized resources over the Internet, capable of providing powerful storage and processing capabilities for big data. Big data is equivalent to the vast amount of knowledge that the human brain memorizes and stores from elementary school to university. Only through digestion, absorption, and reconstruction can this knowledge create greater value. Artificial intelligence can be understood as a process in which robots evolve by absorbing a large amount of human knowledge through continuous deep learning (DL). Artificial intelligence is inseparable from big data, and at the same time, it needs to complete the evolution of DL based on cloud computing platforms.

Cloud computing is like changing from the old model of a single generator to a centralized power model of the power plant. It means that computing power can also be circulated as a commodity, just like gas, water, and electricity, with easy access and low cost. The biggest difference is that its transmission carrier is the Internet. Currently, in the market, there are three types of cloud computing services: infrastructure as a Service (IaaS), Platform as a Service (PaaS), and Software as a Service (SaaS).

IaaS is a service that consumers obtain from a well-established computer infrastructure via the Internet. SaaS is a model of providing software via the Internet, where users do not need to buy software but rent Web-based software from providers to manage business activities. PaaS refers to a platform for software development as a service, which is submitted to users in a SaaS model.

Big data

From the above introduction, big data is a collection of data that is so large that its acquisition, management, and analysis are far beyond the capabilities of traditional database software tools. It is a massive, high-growth, and diversified information asset that requires new processing modes to have stronger decision-making power, insight discovery, and process optimization capabilities. Big data cannot be processed by a single computer, and a distributed architecture must be adopted. Its characteristic lies in the distributed data mining of massive data, but it must rely on the distributed processing, distributed database and cloud storage, and virtualization technology of cloud computing.

In the process of dealing with big data, we first need to collect data, i.e., to obtain all the machine data, log data, and business data through various data sources, followed by filtering and correcting these data to complete data cleaning for the subsequent data processing. For the cleaned data, the third step can be carried out — data modeling, establishing a data storage model, and redesigning and planning the data. Finally, we can complete the data processing, summarize and analyze the various data, and eventually provide a reference for decision-making and application.

If big data is compared to an industry, the key to making this industry profitable is to improve the "processing ability" of data and to realize the "value-add" of data through "processing". It is predicted that from 2016 to 2020, the application value of big data in seven global key fields of education, transportation, consumption, electricity, energy, health care, and finance will be 3.22 trillion to 5.39 trillion US dollars. After experiencing the policy boom and capital boom, the big data industry has entered a steady development stage, and big data has been implemented from the conceptual level to the application level.

Artificial intelligence

At present, artificial intelligence can be divided into three levels: computational intelligence, perceptual intelligence, and cognitive intelligence. Computational intelligence refers to supercomputing power and storage capacity, and AlphaGo is a typical representative. Perceptual intelligence refers to allowing computers to listen, speak, and communicate with people, such as Sophia, the first robot granted citizenship. Cognitive intelligence refers to enabling computers to understand and think, such as various exam robots being developed.

In March 2016, for the man–machine battle between AlphaGo and Sedol Lee, most people still insisted that it is impossible for machines to defeat humans. Even Nie Weiping said bluntly: "If a machine plays a game of Go with a human, I think the machine has no chance at all". However, shockingly, AlphaGo beat Lee by a score of 4:1. The victory of AlphaGo not only set off an upsurge in artificial intelligence but also opened up the evolution of AlphaGo.

After the man–machine battle, in January 2017, an online chess player named Master swept the top players in China, Japan, and South Korea on the Go game website. After winning 60 consecutive victories, he withdrew, and then publicly admitted that he was AlphaGo, and once again received a lot of amazement. Just when everyone thought that Master was invincible, AlphaGo Zero was born. As the latest version of AlphaGo, AlphaGo Zero taught itself for three days and crushed the AlphaGo version that beat Sedol Lee previously with a score of 100:0. After 40 days of self-study, it defeated AlphaGo Master with an absolute advantage of 89:11, once again shocking the world.

Compared with many previous data analysis technologies, artificial intelligence technology is based on neural networks and at the same time develops multi-layer neural networks, which enables deep machine learning. Compared with the traditional algorithm, this algorithm has no redundant assumptions (such as linear modeling needs to assume a linear relationship between the data) but completely uses the input data to simulate and build the corresponding model structure by itself. The characteristics of this algorithm ensure that it is more flexible and has the ability to self-optimize according to different training data.

However, this significant advantage comes with a significant increase in the amount of computation. Before breakthroughs in computation power, such algorithms had limited practical value. About 10 years ago, we tried to use neural networks to compute a set of not-so-massive data, and it might not get the results in three days. But today, high-speed parallel computing, massive data, and more optimized algorithms have jointly contributed to the breakthroughs in the development of artificial intelligence.

The reason why artificial intelligence has attracted much attention in recent years after many years is due to the emergence of DL, a key technology of artificial intelligence, which makes artificial intelligence practical. The concept of DL was proposed by Geoffrey Hinton and others in 2006. It is a new field in machine learning research. It simulates the human brain mechanism to interpret data such as images, sounds, and texts by simulating the neural network that the human brain analyzes and learns. Its essence is to build a machine learning model with multiple hidden layers and learn more useful features through massive training data to ultimately improve the accuracy of classification or prediction.

Same as machine learning methods, deep machine learning methods also include supervised learning and unsupervised learning. Under different learning frameworks, the learning models are also different. For example, Convolutional Neural Networks (CNNs) are a machine learning model under deep supervised learning, while Deep Belief Networks (DBNs) are a machine learning model under unsupervised learning.

DL technology has demonstrated excellent information processing capabilities and can discover more valuable information and knowledge from big data. It has been used in computer vision, speech recognition, natural speech processing, and other fields, and has achieved excellent results. For example, in the medical field, under the training of DL, the accuracy and foresight of image recognition are increasing day by day. Through the processing and objective judgment of patient data, it has reached the level of predicting cancer in advance, and the prediction accuracy is significantly higher than that of human doctors. The Chinese ophthalmologists and scientists have jointly developed a DL algorithm that can identify congenital cataracts, and the diagnostic accuracy rate has exceeded 90%.

It is worth mentioning that artificial intelligence relies on the development of DL, and DL has made substantial progress in the context of the increasingly mature IoT, cloud computing, and big data. By storing the massive data generated and collected by the IoT on the cloud platform, and then through big data analysis, artificial intelligence can provide better services for human production and life. In short, if artificial intelligence is likened to a rocket, then IoT is the launch pad, big data is the fuel, and cloud computing is the engine. Whether it is the IoT, cloud computing, big data, or artificial intelligence, it will become the mainstream of the future market.

Chapter 2

The "Tao" and "Art" of Big Data

From a philosophical point of view, the development of anything is inseparable from the *Tao* and *Art*. The development and application of big data also have inherent *Tao* and *Art*. The *Tao* refers to the beliefs and values that need to be upheld, while the *Art* refers to behavior and methodology. *Sun Tzu's Art of War* made a point that "Tao is the spirit of Art, and Art is the foundation of Tao. We use Tao to guide Art, and use Art to obtain Tao". *Tao* and *Art* are interdependent and work together. If there is only *Tao* but no *Art*, it's just like being an armchair strategist; and if there is only *Art* without *Tao*, it is like a blind man feeling an elephant.

Of course, big data is not a moment of inspiration; it has gone through a period of closely linked development and eventually evolved into the current information asset and scientific paradigm. In this process, it is not only accompanied by the continuous advancement of scientific research but also supported by the development of cloud computing and others. In the end, it appears like a single spark that can start a prairie fire, which has developed from the wild growth to the present gaming era.

The Prototype: The "Fourth Paradigm" of Scientific Research

Driven by the combination of the rapid development of science and technology and the external environment, the paradigm of scientific research is subjected to various new challenges, and new paradigms are constantly emerging. From the development of empirical science to theoretical science, to computational science and data-intensive science, each scientific research paradigm has its own characteristics and paradigms, and has important implications for data engineering and data science. Among them, the influence of the "fourth paradigm" of scientific research is particularly obvious.

What is the fourth paradigm of scientific research?

Before understanding the fourth paradigm, let us first define paradigms precisely. In the view of Thomas S. Kuhn, the author of *The Structure of Scientific Revolutions*, a paradigm refers to the theoretical basis and practical norms by which a scientific community operates in a conventional science. We can understand it as a specific scheme or route that has formed a model and can be directly applied. In short, a paradigm can be understood as a certain norm that must be followed or a routine that everyone uses.

In the field of scientific discovery, thousands of years ago, people mainly recorded and described natural phenomena, which can be regarded as the initial stage of experimental science (or empirical science). Later, we entered the primary stage of scientific development in the Renaissance represented by Galileo. Galileo performed the famous free fall experiment at the Leaning Tower of Pisa, which overthrew Aristotle's statement that "the falling speed of an object is proportional to its weight" and opened a new field for modern science. The experimental-based scientific research model thus established was called the first paradigm (experimental science paradigm).

When experimental conditions were not available, in order to study more precise natural phenomena, a scientific research model based on theoretical research appeared, that was the second paradigm (theoretical science paradigm). The second paradigm was born in the 17th century and lasted until the end of the 19th century. At this stage, scientists used models to simplify the scientific principles that could not be simulated experimentally, removed some complex factors, left only the key ones, and then draw conclusions through algorithms. For example, we are familiar with Newton's first law: all objects remain at rest or move in a straight line at a uniform speed when they are not acted upon by a force.

At the end of the 19th century, the second paradigm developed to the extreme. At that time, Newton's three laws explained classical mechanics, and Maxwell's theory explained electromagnetism, which made the grand edifice of classical physics at that time. At the beginning of the 20th century, the two mountains of quantum mechanics and relativity rose again, opening another golden age of science. Both quantum mechanics and relativity were coincidentally based on theoretical research when extraordinary minds and calculations transcended experiments.

With the increasing difficulty and economic investment in verifying theories, scientific research was gradually becoming exhausted. At this point, another scientist, John von Neumann, stepped forward. He proposed the architecture of modern electronic computers in the mid-20th century, which continues to this day. Since then, with the rapid development of electronic computers, the third paradigm (computational science paradigm), which uses electronic computers to simulate scientific experiments, has been rapidly popularized. Whether in basic science research or engineering experiments, computer simulation is increasingly replacing experiments and becoming a common method in scientific research.

Computational science has a wide range of applications, including numerical simulations to reconstruct and understand known events, such as earthquakes and tsunamis, or to predict future conditions, such as weather forecasts. It can also be used to perform model fitting and data analysis, adjust models, or use observations to solve equations for oil exploration. In the fields of crafts and manufacturing processes, computational and mathematical optimizations are performed to find optimal solutions.

Since then, with the advent of the Internet age, Jim Gray, a Turing Award winner, in view of the explosive growth of data, believes that the data-intensive paradigm should and has been separated from the third paradigm (paradigm of computational science) and become a unique scientific research paradigm, that is, the fourth paradigm (data-intensive paradigm). Since then, based on the essence of his last talk, Gray's colleague Tony Hey and others have written a book named *The Fourth Paradigm: Data-Intensive Scientific Discovery*.

Characteristics of the fourth paradigm

The data-intensive paradigm is the fourth paradigm after experimental science, theoretical science, and computational science. Compared with other paradigms, this paradigm has its special features.

In the era of big data, people are more concerned about the correlation of things than desiring for causality. In this context, as a scientific research method tailored for big data, the most distinctive feature of the data-intensive paradigm lies in its *objectivity* and the pursuit of *relevance*. If we consider *causality* as a special kind of correlation, then big data is actually a more generalized *correlation*.

Compared to the computational science paradigm, the data-intensive paradigm is also computed by computer, so what is the difference between the two? For the computational science paradigm, theoretical assumptions are often made first, and on this basis, calculations and verifications are performed by collecting data and materials, while the data-intensive paradigm is based on a large amount of data and the unknown conclusions are directly drawn through calculations.

As Jim Gray interprets, the methodological procedures of the fourth paradigm are accomplished through the following steps: (1) acquisition of data through tools or simulations, (2) software processing, (3) storage of information or knowledge in computers, and (4) analysis of databases and documents by scientists through data management and statistics.

In the fourth paradigm, scientists need to solve hard-to-reach scientific problems by monitoring data in real time and dynamically. In the process, the role of data has also shifted. Data are no longer just the result of scientific research but also becomes the object and tool of scientific research, and the basis of scientific research. Scientists need to think, design, and implement scientific research based on data.

Specifically, people no longer only care about data modeling, preservation, analysis, reuse, etc. but explore data, and its inherent interactivity, openness, and knowledgeable objectification and computability based on massive data have become the key points for scientific research to focus on. In this process, people construct a data-based, open, and collaborative research and innovation model, and conduct fourth-paradigm scientific research, that is, data-intensive knowledge discovery.

Of course, this is due to the rapid development of information and network technology. Through the deployment of infrastructures such as perception, computing, emulation, and simulation, it is becoming easier to obtain and study data. This also provides more possibilities for practical research and application while further promoting the development of the fourth paradigm. For example, in terms of environmental monitoring and governance, diverse environmental data are now extensively collected through various types of sensors that can locate spatial locations. Based on its research and analysis, further pollution tracing, early warning, and even pollution control can be accomplished.

As we can see above, the fourth paradigm is to let computers discover patterns from massive data and let the data speak for themselves, and the truth lies in the "data". As a new channel of knowledge discovery, the fourth paradigm should be recognized. Its appearance is not to deny the first three paradigms but to complement the first three paradigms, and they together constitute a method system for people to discover knowledge and search for truth.

Breaking Out of the Shell: McKinsey Predicts the Arrival of "Big Data"

Big data is a concept that continues to evolve. With the development of IT technology and the continuous accumulation of data, big data technology begins to be constantly mentioned. In 2008, *Nature* magazine published an article, "Big data: The next Google". The article begins with "Ten years ago this month, Google's first employee turned up at the garage where the search engine was originally housed. What technology at a similarly early stage today will have changed our world as much by 2018?"

The article lists the views of many experts, researchers, and business people. Although it involves a wide range of fields, everyone agrees that breaking the boundary between virtual and reality, integrating material and information in the world, and creating a huge database will be an important change in the future. In other words, *Nature* lists big data as an innovative change comparable to Google's search engine in the next 10 years.

Since the new term "big data" was coined by *Nature* in 2008, it has appeared frequently in the public eye. In 2011, *Science* magazine pointed out in its special issue *Special Online Collection: Dealing with Data* that by collaborating with colleagues in many scientific careers, *Science* has summarized the challenges brought by the flood of data and the opportunities brought by the rational organization and utilization of data in response to the growing research data.

Although both *Nature* and *Science* have introduced the term "big data" to the public as a cutting-edge technology vocabulary, the definition of big data has not been clearly defined until the McKinsey Global Institute (MGI) officially introduced it into the concept definition stage.

McKinsey was the first to put forward the view that "the era of big data has arrived".

In 2011, McKinsey pointed out in the research report *Big Data: The Next Frontier for Innovation, Competition, and Productivity* that data had penetrated into every industry and business function area, gradually becoming an important production factor, and the use of massive amounts of data would herald a new wave of productivity growth and consumer surpluses.

In the report, McKinsey defines "big data" as data sets whose size exceeds the capabilities of typical database software to collect, store, manage, and analyze. This definition has two connotations: First, the size of datasets that meet big data standards will grow over time and technological progress; second, there are differences in the size of datasets that meet big data standards in different departments. Currently, the general range for big data is from a few terabytes to several petabytes (thousands of terabytes).

As technology advances, the size of datasets considered "big data" will increase by orders of magnitude. At the same time, the definition of the size of a dataset that is considered "big data" varies by sector, depending on the software tools commonly used in those sectors and the typical size of the dataset.

Coincidentally, for the development of big data, Gartner, a well-known international consulting agency, also regarded 2011 as the technological budding period of big data in its emerging technology maturity curve. At the same time, Gartner also regarded 2012–2014 as a period of technical hype for big data. After crossing the hype period, big data gradually cooled down and began to enter the substantive data research stage. In 2013, big data turned from a technical buzzword into a leading social wave that gradually began to influence all aspects of social life. In just a few years, big data has evolved from a technical term identified by large Internet companies to a major technology proposition that will determine how we live digitally in the future. At the same time, AI is also setting off the next technological wave.

Rise: Cloud Computing Creates Technical Prerequisites for Big Data

Big data and cloud computing complement each other. On the one hand, cloud computing provides powerful storage and computing capabilities for big data, as well as higher-speed data processing, which provided services more conveniently. On the other hand, business needs from big data will find more and better practical applications for the implementation of cloud computing. In general, cloud computing creates the technical prerequisites for the rise and development of big data. If big data is compared to a resource-rich mining pool, then cloud computing is the tool and approach to gold nuggets.

After 10 years of development, cloud computing has reached a scale of over 10 billion in China. It is no longer just a tool for storage and computing but has been widely used and become a mainstream technology and business model in the IT industry. We can think of the cloud as an apple tree full of big data.

Briefly speaking, cloud computing is an Internet-based computing method through which shared software and hardware resources and information can flow to computers and other devices on demand, like water and electricity. At the same time, the National Institute of Standards and Technology (NIST) defines five basic characteristics (self-service on demand, extensive network access, resource sharing, fast availability scalability, and metered pay services), three service models, and four deployment methods of the cloud computing model, which can provide flexible and fast resource services.

Among them, cloud computing has three basic service models: SaaS, PaaS, and IaaS. Among them, IaaS is a cloud computing service provider renting out processing, storage, networking, and other basic computing resources, and consumers obtain services from a complete computer infrastructure through the Internet. PaaS is to deploy applications created or acquired by consumers to the cloud infrastructure using programming languages and tools specified by resource providers. SaaS is where applications run on cloud infrastructure, and consumers do not directly manage or control the underlying infrastructure.

Simply speaking, the three service models can be thought of as the process of buying a pizza. If you want to eat pizza, then you can choose to buy instant pizza to bake at home, that is, you need a pizza supplier (IaaS) that provides materials. Or, if you think this method is a bit troublesome, you can also place an order through a takeaway platform, and the pizza will be delivered to your door in a short time (PaaS). Of course, if you happen to be close to the pizzeria, you can also go directly to the restaurant to eat. Compared with the first two methods, this is simpler and more straightforward, because even the dining table and paper towels are ready-made, and all you need to do is wait to eat (SaaS).

In addition to the three service models, cloud computing also has four deployment modes: private cloud, community cloud, public cloud, and hybrid cloud. Among them, a private cloud is built for a single user to use alone. A community cloud is a cloud created by organizations that share common interests and intend to share infrastructure. A public cloud is publicly available to the public or industry organizations. A hybrid cloud is a combination of two or more abovementioned clouds.

Through the introduction of the characteristics and modes of cloud computing, it can be noted that cloud computing has the characteristics of elastic scaling, dynamic allocation, resource virtualization, support for multi-tenancy, support for metered billing or on-demand use, etc., which just fits the development needs of big data technology. With the explosive growth of data volume and higher and higher requirements for data processing, only the combination of big data and cloud computing can better utilize the advantages of both.

At present, big data is gradually playing a pivotal role in many aspects such as precise marketing, intelligent logistics, intelligent transportation, and safe cities. It is worth mentioning that big data is also showing the application trend of combining with AI, where computers can better learn to simulate human intelligence. For example, new progress has been made in speech recognition, machine translation, etc., and machine learning is also playing an important role in more and more fields.

Frontier: From the Wild Growth to the Era of Gaming

The biggest energy source in the future is not oil but big data. Whether in terms of the global environment or just talking about the individual cases of big data development in China, big data has shown a trend of explosive growth. According to the international data information company, it is estimated that the total global data volume will reach 44 ZB in 2020, and the Chinese data volume will reach 8,060 EB, accounting for 18% of the total global data volume. From finance to medical care, from advertising to e-commerce, the thirst for data in all walks of life is unprecedented.

Corresponding to the explosive demand, big data companies began to grow wildly. This change can also be clearly found in the big data ecological map made by Matt Turck, former managing director of Bloomberg Ventures and partner of FirstMark Capital, in recent years.

In 2014, although the development of big data was slower than it is now, it has begun to develop towards the application level. Hadoop has become a key part of the infrastructure, and Spark has also begun to become a complementary framework. From the big data ecological map of the year, it can be seen that more and more enterprises are joining this emerging field.

The current state of the increasingly crowded big data industry in 2014. Through the industry map in 2017, it is easy to find that at this time, the hype of big data further dissipated and entered the deployment state. More players flocked to it and the applications became more diversified. Meanwhile, the year showed some new trends that cannot be ignored.

(1) Big data companies, which were startups in previous years, have grown in strength after just a few years of development. They have been listed one after another, and many super unicorns have emerged. Meanwhile, for the new influx of big data companies, mergers and acquisitions have become commonplace and have continued steadily.

(2) As AI becomes a new outlet, there are more and more vertical applications driven by AI technology, and big data and AI are more and more closely integrated. "Big data + AI" is becoming a technology stack for modern applications, where big data is the pipeline and AI provides intelligence.

(3) For cloud services, "functional integration" has become a trend. The internationally renowned cloud providers, such as AWS, are improving cloud services in combination with current development hotspots. AWS has analytical frameworks, real-time analytics, databases (NoSQL, graphs, etc.), business intelligence, and an increasingly rich set of AI capabilities, encompassing almost all services in big data and AI.

(4) The data-driven world is followed by the game of data openness and security. Even though big data can solve many problems in people's life and development, it does not mean that personal privacy, corporate information, and/or even national security need not be protected. On the contrary, in the era of big data, it is more demanding to protect the information of individuals and enterprises so that the technology of the information age can benefit each individual to the greatest extent.

The first and foremost issue is the privacy of individual users. In the era of big data, personal information that people fill in on social platforms and the location information and health information uploaded through various sports equipment can be easily collected by service providers to build a high-precision personal information system. One example is the recently exposed Facebook data breach incident. According to foreign media reports, more than 50 million Facebook users' information data were leaked through Cambridge Analytica, which used algorithms to carry out "precision marketing" business operations to users, thus affecting the results of the 2016 US election and causing worldwide outbreaks.

Moreover, for enterprise information security, whether it is due to negligence or improper management leading to corporate data leakage, intentional hacker attacks, or virus Trojan horses, it will inevitably have a significant impact on the security, brand, and image of the company, and even directly decide the life and death of the enterprise. This is a big challenge for big data companies. At the same time, for big data businesses, traditional security protection concepts have certain limitations, and more effective methods are urgently needed to deal with hacker attacks or internal security control.

Finally, data security is far more than a trivial matter that individuals or enterprises need to worry about. From the national level, there are also hidden dangers of data security. In particular, for important institutions such as information facilities and military agencies, the security of data and information is directly related to the security of the country's technology, politics, military, culture, economy, and other aspects. It is safe to say that data security is a sensitive area that is related to the national economy and people's livelihood.

From "Data" to "Intelligence"

In the era of big data, the amount of data in all walks of life has grown exponentially, but most of these rapidly expanding data are fragmented and unstructured data. For these samples of data, if there is a lack of effective processing and refining, it is actually just invalid information and cannot provide any valuable reference for decision-making. Therefore, in the process of data collection, processing, and application, how to extract data into knowledge, and then develop it into wisdom, to remove the false and extract the truth, has become an important part of knowing the "Tao" and "Art" of big data.

The development from "data" to "wisdom" is not achieved overnight. Before discussing its realization process, we need to clearly discuss the starting point of the problem, and it is also the key concept that we should really care about. For example, we need to clarify four concepts, namely *data*, *information*, *knowledge*, and *wisdom*, which together form a basic model: the DIKW (Data–Information–Knowledge–Wisdom) hierarchical model (Figure 2.1).

Data are recorded and identifiable symbols of objective things, including numbers, words, graphics, audio, and video. It is the original material of the DIKW process and provides samples for data processing. Information is the processing of data to establish connections with each other, making it meaningful and usable data. Knowledge is the further processing and analysis of information and its internal connections, from which the required regular understanding is obtained. It is the application of information. Wisdom is based on existing knowledge and advanced comprehensive ability, discovering the principles, predicting the development of objective things, etc. It is the application of knowledge.

For example, a teacher has obtained a mid-term test report card of the class, which records the scores of each subject and the total scores of each student in the mid-term test. These individual grade data were initially just some scattered raw "data" — scores xxx. After being analyzed and sorted, the data gradually turned into "information" — the average score of the class was 507 points. At this time, through the comparative analysis and grade ranking with the previous test scores, the "knowledge" that "the grade ranking of this class has risen by three places in this test" is further obtained. Through excluding other influencing factors, this knowledge suggests that the learning status and learning methods of the classmates in the recent period are good. Eventually, the teacher obtains "wisdom" that is to continue the past teaching methods, which provides a reference for decision-making.

Specifically, it can be divided into multiple stages such as from data to information, from data to knowledge, and from data to wisdom (Figure 2.2).

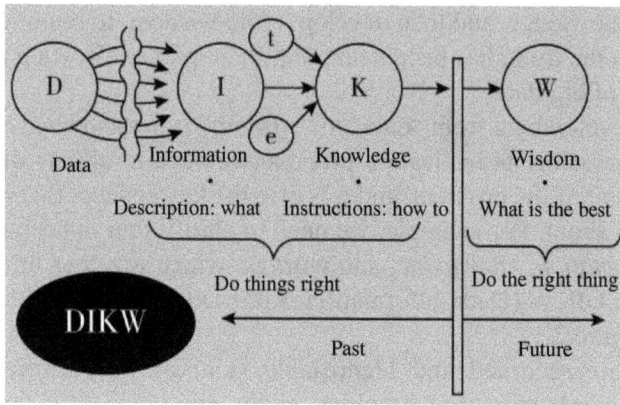

Figure 2.1   DIKW Model Flow Chart

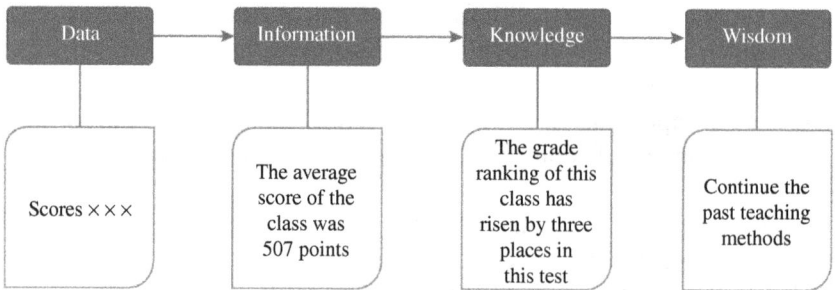

Figure 2.2   Data–Information–Knowledge–Wisdom Transformation Process

From data to information

Data can exist in various forms such as tables, images, and text. In the DIKW model, data only represents the data itself and does not contain any underlying meaning. In the medical industry, such as in a hospital in Toronto, Canada, more than 3,000 data readings per second (including heart rate, respiration, temperature, and blood pressure) are performed on premature babies in order to better analyze patient information. Without analyzing the data, these individual data alone are meaningless.

However, if these data are organized and processed in a certain way, the data will have meaning. Questions such as *Why, What, Where,* and *When* can now be answered because data are converted to information. At this hospital, after data collection and analysis on premature babies, doctors will know which babies may have problems and take targeted measures to prevent them from dying.

From information to knowledge

Knowledge is the collection of information that makes it useful. Knowledge is the application of information. It is a process of judging and validating information that combines experience, context, interpretation, and reflection. Knowledge can answer the questions of *how* and can help us with modeling and simulation.

Knowledge is useful information that is filtered, refined, and processed from relevant data. In a particular context, knowledge creates meaningful connections between data and information, and information and the application of the information in action. It embodies the nature, principles, and experience of information.

For knowledge, we need not only simple accumulation but also understanding. Comprehension is a process of inference and probabilism, a process of cognition and analysis, and it is also a process of creating new knowledge based on the information and knowledge already grasped. At the same time, knowledge is based on reasoning and analysis, and it may also generate new knowledge.

Based on the data collected from preterm infants, correlation analysis is performed so that valuable knowledge can be obtained to determine which infants are more prone to emergencies. Based on data and information, combined with experience and background information (such as the doctor's past diagnostic experience and the baby's status at birth), the vital signs of premature infants can be determined, and any spot signs of infection could be estimated 24 hours in advance so that measures can be taken to safeguard the health of premature babies.

From Knowledge to Wisdom

Wisdom is a unique ability displayed by human beings, which is mainly manifested in the ability to collect, process, apply, and disseminate knowledge, as well as a forward-looking view of the development of things. On the basis of knowledge, through the accumulation of experience and insight, the profound understanding and foresight of things formed, which is reflected as a kind of excellent judgment. Unlike the previous stages, wisdom focuses on the future, trying to understand what was not understood in the past, and what was not done in the past. It is unique to humans, and it is the only thing that cannot be achieved with tools.

Wisdom can be summed up simply as the ability to make sound judgments and decisions, including the optimal use of knowledge. Wisdom can answer the *why* question. Going back to the previous example, when collecting data on preterm infants, we only need to consider how to get more precise information about each baby through the sensing device. However, when conducting predictive work such as monitoring premature infants and nursing improvement plans, it is necessary to further consider

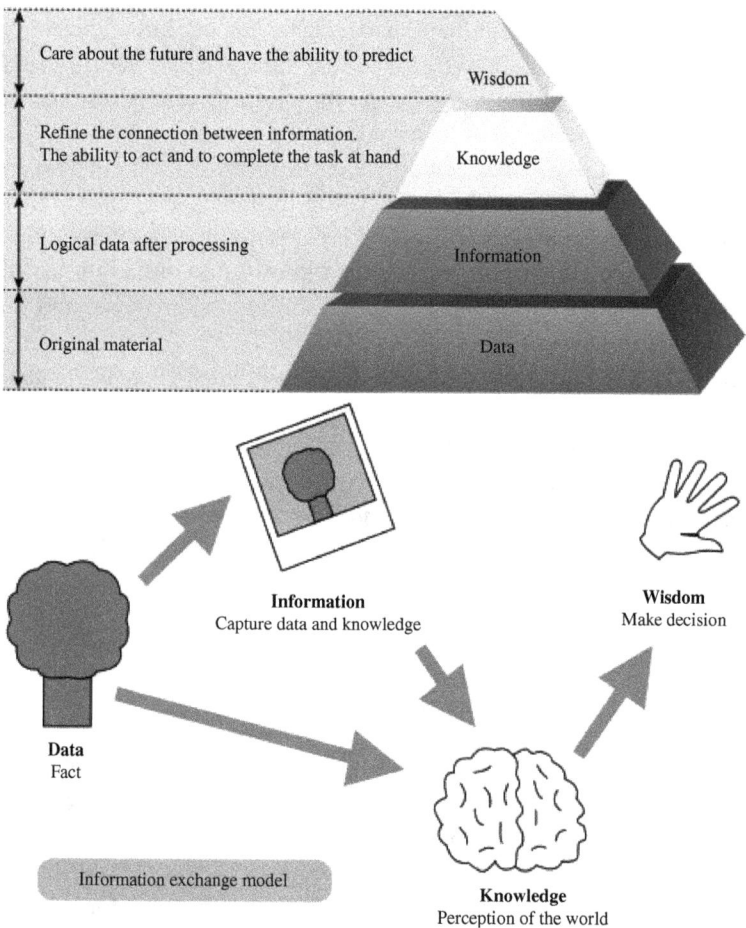

Figure 2.3   The Relationship between Data, Information, Knowledge, and Wisdom

the overall data situation of a certain area and a certain period. That is, the formation of predictive intelligence based on data, information, and knowledge (Figure 2.3).

"Extensive Reading" and "Profound Thinking"

Big data is everywhere: it affects our work, life, and study all the time, and will continue to have an even greater impact. Big data can transform production and lifestyles: on the one hand, it is inseparable from its wide range of sources; on the other hand, with the proliferation of data, the amount of data gradually accumulates to a critical point. Once this critical point is exceeded, the underlying laws and values will gradually appear. Consequently, the future development trend can be captured. In this process, the data needs to be "big enough".

For data, only correlation and analysis can bring value. With the explosive "growth" of data, the magnitude of big data is no longer a problem, but how to conduct in-depth processing and analysis based on this "big" data has become an important issue. In this process, we realize the analysis and processing of extract–transform–load (ETL) to achieve deep thinking, and then derive information and acquire knowledge, and eventually transform data into wisdom.

For example, Target, a mass consumer supermarket in the United States, has once obtained the purchase records of all users based on more than 20 products that women may purchase during pregnancy. Based on this data source, a model is constructed to analyze the behavioral correlation of buyers, so as to accurately infer the specific delivery time of pregnant women. According to Target's data analysis, expectant mothers are prone to buying some unscented lotions when they are three months pregnant. A few months later, they will need to further supplement calcium, magnesium, zinc, and other nutrients. Based on these results, Target's sales department can send corresponding product coupons to each pregnant customer at different stages, thereby increasing the sales rate.

For another example, for brand shoe stores such as Adidas, which have many stores around the world, their strategies on how to quickly and effectively coordinate their stores, accurately grasp market trends, and realize the matching of their own products with user needs also rely on big data.

From a macro perspective, consumers in first- and second-tier cities pay more attention to fashion and branding, while consumers in third- and fourth-tier cities pay more attention to practicality. Adidas will analyze and mine based on the data provided by dealers, and then provide more valuable sales suggestions. For example, they could tell the dealer in the lower-tier market that products in this color are more likely to be sold out because local consumers prefer blue.

Now, Adidas' "let data speak" method can help dealers choose more suitable products and put products that meet consumers' needs more accurately into the corresponding regional markets. It is these objective facts based on data discovery that help Adidas formulate a more suitable sales strategy for dealers. While promoting the hot sale of products and increasing the sell-out rate, it avoids the occurrence of problems such as excessive inventory.

Empowering Big Data with Big Wisdom

The value of big data is beyond all doubt. What we need to discuss is how to make better use of big data. At present, there are still many problems in this regard.

One is the challenge of the scientific basis. For big data, only analysis and correlation can bring value. Unlike traditional statistics, which selects abstract data as the object of analysis, big data processing deals with natural data. The limiting distribution is not stable and is not suitable for use with traditional sampling mechanisms. Therefore, as far as the scientific basis of big data is concerned, it is necessary to rebuild the foundation and establish the basic theoretical system of data science. According to Viktor Mayer-Schönberger, it can be considered as three changes in the way of thinking: (1) To analyze all the data related to something, instead of relying on analyzing a small number of data samples. (2) We are willing to accept the complexity of data, and no longer pursue accuracy. (3) Our way of thinking has changed. We no longer look for elusive cause-and-effect relationships but focus on correlations.

The second is the challenge of computing technology. In addition to the scientific basis, big data applications also depend on changes in core technologies, such as storage computing architecture, query and processing computing modes, and various programming languages and algorithms for computing, analysis, and mining. All these must be reformed accordingly.

The third is the challenge of authenticity determination. Big data produces great value, but in practical applications, it is difficult to guarantee that every "big value" is correct. Whether it is the source of data or the method of decision-making, if it is used improperly, big data can inevitably turn "big fool". Therefore, authenticity determination remains to be a big challenge.

Therefore, in the process of big data processing and application, it is urgent to endow big data with "big wisdom". In addition to laying the applicable scientific foundation, updating the corresponding computing technology, and checking the authenticity, AI cannot be ignored. Intelligence is neither sourceless water nor rootless tree. If AI aims to allow machines to acquire intelligence, it must also be based on massive data cornerstones and conduct continuous training and learning to form wisdom. In a certain sense, big data supports the realization of AI, and AI is the goal of big data applications.

Data mining (DM), for example, is a typical application of AI processing data. DM originated from knowledge discovery in databases (KDD), which can be simply understood as the process of searching for information hidden in a large amount of data through algorithms.

In fact, before the emergence of AI, data processing was mainly composed of data collection, data conversion, data grouping, data sorting, data calculation, data storage, data transmission, data retrieval, and other parts. It is often time-consuming and laborious to extract and derive valuable and meaningful data for some specific groups of people.

In the AI era, the application of AI technology, supported by the three key points of powerful computing power, efficient algorithms, and sufficient and large amounts of training data, can greatly reduce the consumption of manpower and material resources, and complete data processing work quickly and efficiently. For example, for a driverless car, its own GPS, RFID, sensors, etc. become the *eyes* and *ears* of the car, which can collect, store, and send all the working conditions and static and dynamic information of the vehicle. Whenever signal data such as obstacles in motion are collected and processed by AI technology, the *brain* is able to direct the car to drive accurately along the road. It only takes a few seconds from data collection, calculation, and analysis to the final result output, and no one is involved in the whole process.

The core of AI's intelligence lies in machine learning and deep learning based on feeding massive data. At the same time, the more data you have, the more efficient the neural network becomes. That is to say, as the amount of data grows, so does the number of problems that machine language can solve. For example, researchers from the Department of

Radiology at Massachusetts General Hospital and Harvard Medical School used convolutional neural networks to identify CT images. They evaluated the accuracy of the neural networks based on the size of the training data. As the training size increases, the accuracy will improve.

Big data and AI complement each other. Based on the analysis and processing of massive data, AI is becoming more and more "smart". In turn, intelligent AI can conduct in-depth analysis and processing of data to draw more valuable and accurate conclusions, provide references for management decisions, and make big data with great wisdom.

Based on this, China has identified the relationship between big data and AI, and has focused on building its AI first-mover advantage. The *New Generation Artificial Intelligence Development Plan* takes "big data intelligence" as one of the five important directions of planning and deployment. The methodology of "big data + AI" for intelligent application of big data. It uses AI to conduct an in-depth analysis of big data, explore the intelligent form of its implicit patterns and laws, and realize theoretical methods and supporting technologies from big data to knowledge, and then to decision-making wisdom. Finally, a general AI model will be established to release the "intelligence bonus" of big data intelligence.

Chapter 3

Big Data Processing System

Big data is bringing about a revolution in the information society. The widespread application of large amounts of structured and unstructured data makes people rethink the existing IT model. At the same time, big data is promoting another business transformation based on the information revolution, enabling society to obtain more social benefits and development opportunities with the help of big data. Therefore, the related technologies of big data processing have also received great attention from all walks of life.

This chapter starts with the development history of big data processing, systematically introduces the basic concepts of big data processing, and analyzes the current big data processing mode for different data processing environments. From a technical point of view, this chapter further explains in detail the architecture of typical big data processing systems such as Hadoop and Spark, and guides readers to understand big data storage, big data processing, and finally to understand the application of big data processing in various industries through specific examples.

Basic Concepts of Big Data Processing

Speaking of big data processing, it all started with Google's classic paper: *MapReduce: Simplified[1] Data Processing on Large Clusters*. At that time, due to the sharp increase in the number of web pages, Google usually had to write a lot of programs to process a large amount of raw data, such as

---

[1] In the original text: simplified.

web page data scraped by crawlers, web page request log data, and various types of derived data. These calculations are conceptually easy to understand. However, due to the large amount of input data, it is impossible for a single machine to complete the processing. Therefore, we need to complete the calculation in a distributed manner. In this process, it is necessary to consider how to perform parallel computing, distribute data, and deal with failures.

In response to these complex problems, Google has designed the MapReduce computing framework to achieve large-scale parallel computing. The biggest contribution of MapReduce to big data processing is that this computing framework can run on a group of inexpensive servers. It has changed the public's understanding of big data computing and transformed big data computing from centralized computing to distributed computing. In the past, if we wanted to calculate more data, we had to build better and faster computers, but now we only need to add server nodes. Since then, the historical new chapter of big data processing has opened.

The data in big data is generally divided into two types: one is called structured data, and the other is called unstructured data. Structured data is data with a fixed format and limited length, such as nationality, ethnicity, name, gender, and data in tables. The amount of unstructured data is relatively large and with no fixed format and/or length, such as video data and voice data. Looking at the data alone is meaningless. It must undergo certain processing so that we can sort out the key information from the messy data. Usually, this information contains many rules. We need to summarize the rules from the information to form "knowledge", which is the value of data.

For example, Google built a specific mathematical model by analyzing 50 million frequently searched words by Americans and comparing them with data from the United States Centers for Disease Control and Prevention during the 2003–2008 seasonal influenza spreading. Based on the model, Google successfully predicted the spread of flu in the winter of 2009, even down to specific regions and states. Another example is Ali Small Loan, which can lend money to customers without meeting them through an all-round evaluation of downstream orders, upstream suppliers, and business credit of loan customers. The data come from a large data-sharing platform, which creates business value by sharing the data resources of Alibaba's subsidiaries. The big data team integrates and interconnects the data of each link of the Taobao transaction process, and then classifies, stores, analyzes, and processes the information based on business understanding, and finally analyzes and mines the results together with the decision-making behavior.

It can be seen that the above two cases involve massive amounts of data. It is unrealistic to extract important information from these massive amounts of data and realize the value of big data without sufficient computing resources. So, how to deal with the huge amount of data? Speaking of which, everyone should think of cloud computing.

At the inaugural meeting of the CCF Task Force on Big Data, Academician Huai Jinpeng, the director of the committee, used a formula to describe the connection between big data and cloud computing: $G = f(x)$, where x is big data, f is cloud computing, and G is our goal. In other words, cloud computing is a means of processing big data, and big data and cloud computing are two sides of the same coin. Simply speaking, big data is the demand, and cloud computing is the means. Without big data, cloud computing is unnecessary. Without cloud computing, big data cannot be processed.

Big Data Processing

Massive data requires us to perform operations, such as stripping, sorting, classification, modeling, and analysis. After these operations, we start to establish the data analysis dimensions, that is, analyze data of different dimensions, and finally get the desired data and information. Big data processing technology is a technology that quickly obtains valuable information from various types of data. The processing generally includes big data collection, big data pre-processing, big data storage and management, big data analysis and mining, and big data display and application.

Big data collection

To process big data, we must first have data. The three main ways we commonly use to collect data are as follows:

(1) *Data capture*: Extract relevant information from existing network resources through programs and then record it into the database. It can be roughly divided into URL crawling and content crawling methods. URL crawling is to quickly crawl the required URL information through the setting of URL crawling rules. Content crawling is to accurately capture the scattered content data in the web page by

analyzing the source code of the web page, and setting content crawling rules. The content crawling can be completed on complex pages, such as multi-level and multi-page.

(2) *Data import*: Import the specified data source into the database. The supported data sources include databases (such as SQL Server, Oracle, MySQL, and Access), database files, Excel tables, XML documents, and text files.

(3) *Automatic information collection of IoT sensing devices*: IoT sensing devices are functionally composed of power modules, acquisition modules, and communication modules. The sensor transmits the collected electrical signal to the main control board through the wire. The main control board performs the signal analysis, algorithm analysis, and data quantification, and finally, the processed data is transmitted through general packet radio services (GPRS).

Big data pre-processing

The data in the real world are generally incomplete and inconsistent rough data, which cannot be directly mined, or the mined results are unsatisfactory. In order to improve the quality of data mining, data pre-processing techniques have been developed. There are various methods for data pre-processing, including data cleaning, data integration, data transformation, and data reduction. Data pre-processing is mainly used to complete the extraction and cleaning of the collected data, which greatly improves the quality of data mining and reduces the time required for data mining.

(1) *Extraction*: Because the acquired data may have multiple structures and types, the data extraction process can help us convert these complex data into a single or easy-to-process structure and type, so as to achieve the purpose of quick analysis and processing.

(2) *Cleaning*: For big data, it is not all valuable, some data are not what we care about, while others are completely wrong interfering items. Therefore, we need to filter the data to extract valid data.

Big data storage and management

Big data storage and management are to store the collected data, establish the corresponding database, and manage and call it.

Today, the continuous growth of data can no longer be dealt with by a single-machine system alone. Moreover, in the face of the growing data, how to do a good job of big data storage is the core issue of big data technology. Even if the configuration of hardware equipment is continuously improving, it is still difficult to keep up with the speed of data growth. Fortunately, the current mainstream computer servers are relatively cheap and have excellent scalability. For now, buying 8 servers with 8-core processors and 128 GB of memory today is far more cost-effective than purchasing a server with 64-core processors and terabytes of memory, and you can also increase or decrease servers to deal with future data storage pressures. This distributed architecture strategy is more suitable for the storage of massive data. At present, common big data storage technologies include distributed database systems represented by NoSQL databases, distributed file systems represented by Hadoop distributed file system HDFS, and DRAM-based in-memory data management technologies.

Big data analysis and mining

The collected data are raw data. Most of the raw data are disorganized and contain a lot of useless data. Therefore, data pre-processing is required to obtain some high-quality data for storage. For high-quality data, data analysis can be performed to classify the data or discover the interrelationships between the data to obtain valuable information. For example, the story of beer and diapers in Walmart supermarket is based on the analysis of people's purchase data, through which we find that men usually buy beer when they buy diapers. In this way, the relationship between beer and diapers is drawn. Walmart put the results of data analysis into practice and get the rewards.

Speaking of data mining, we have to mention data retrieval. The so-called retrieval is to search. The two major search engines, Google and Baidu, put the analyzed data into the search engine. When users want to find information, they can search for it. However, just getting information through searching can no longer meet the users' needs. It is also necessary to dig out the mutual relationship from the information. For example, stock securities search. When searching for a company's stock, is it suitable to also mine the company's executive information? What if the shareholders buy because they only get the information that the company's stock is rising, yet later the executives outbreak negative news and cause the stock to fall? In this case, the partial information of rising stock defrauds the shareholders. Therefore, it is very important to mine the relationship in the data through various algorithms and form a knowledge base.

Big data display and application

Big data processing technology can dig out the information and knowledge hidden in the massive data, and provide the basis for people's social and economic activities, thereby improving the operation efficiency of various fields and intensification of the entire social economy. China is at the forefront of the world in the display and application of big data, especially in the fields of business intelligence, government decision-making, and public services. For example, the rapid popularization of Internet finance and credit system products based on big data, the management and application of smart traffic based on big data, and the matching of cost reduction and efficiency improvement by providing real-time information and data for cargo owners and drivers in smart logistics. At the same time, with the formulation and implementation of supporting policies and measures of the national big data strategy, the development environment of China's big data industry is constantly being optimized. New formats, new businesses, and new services of big data will emerge with explosive growth, and the industry chain will also be further mature and expanded.

Big Data Processing Model

In fact, the differences between big data processing and traditional data processing are not so great. The main difference is that big data has to deal with a massive amount of unstructured data, so parallel processing can be used in all processing stages. According to the classification of the processed data form and the timeliness of the obtained results, big data processing models can be divided into four types: relational model, graph model, stream model, and parallel decomposition model.

Relational model

The relational model refers to the relational database query through the relationship, that is, by establishing the index relationship of keywords, and finally finding the set target. For example, on the premise of knowing a person's ID number, the person's address can be found through the fast association of the ID number through the population database. At the same time, through the ID number, it is also possible to search for the person's orders to buy air tickets or book hotels in various places. This is a typical relational model. Traditional databases are relational models.

Since the relational database has a strong indexing relationship, the information can be found quickly according to the keyword index during searching, with no need to search every number. Thus, the search is fast. For example, when searching for ID numbers, the relational database sorts them. If you search 10 times, you can find 2^{10} people, that is, 1 billion people. By binary search, you can quickly get the search results. In other words, if you look up the ID numbers of people in the whole country (about 1.3 billion people), you can find them in 11 searches. Because only sorting is needed, the search is fast, which is the benefit of relational data-bases. However, due to the strictness of this model, data must be regular-ized, structured, and strictly organized, which is the disadvantage of the relational data model.

At present, a lot of data are unstructured. For example, because the Internet is unstructured, a lot of information on various web pages is unstructured or semi-structured. It's harder to deal with such data. Among them, video data is typical unstructured data. It has no structure and is all images. The premise of processing these data is to turn them into struc-tured data.

For such data, the video needs to be identified first, i.e., identify a specific face and use the ID number to represent it in a certain place at a certain time, thereby converting unstructured data into structured data. In the process of city-level video application, the video is often indexed dur-ing the recording process, and the structuring work is carried out simulta-neously. After that, the target person is indexed by the ID number, and the original video where this person appears can be easily found in the next search.

Graph model

In life, when we need to express the interrelationship between objective affairs, especially the intricate interrelationship between multiple things, the language description is often unsatisfactory, and sometimes the more the description is, the more complex it is. At this time, a relationship

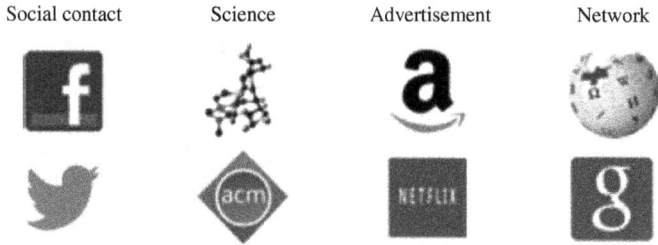

Figure 3.1   Application Fields of Graph Computing

diagram can clearly express the relationship between them. The same goes for data relationships. Based on "graph theory", the abstract expression of a "graph" structure can be made for data. Consequently, graph computing emerges (Figure 3.1). Graph computing is presented in the form of graphs or networks, such as social networks, infectious disease transmission routes, and road traffic flow. It clearly represents the relationship between objectives through intuitive and clear image expression.

In short, graph computing technology is a technology to describe, calculate, and analyze the relationship between things. A graph is a data structure composed of vertices with attributes and edges with attributes. Specifically, it operates based on the three major capabilities of artificial intelligence: understanding ability, reasoning ability, and learning ability, namely URL.

For understanding ability, through graph computing technology, the interrelationship between things can be completely expressed. For reasoning ability, sometimes the interrelationships between things are not obvious, but they can be quickly deduced through the graph. For learning ability, in the dynamic and highly correlated data world, through graph computing, objectives can be summarized, deduced, and described, so as to abstract and sublime, and finally learn and improve.

In practical applications, graph computing is often used to express the relationship in three aspects: (1) human–person relationship, (2) human–behavior relationship, and (3) human–things relationship. The following describes the above three relationships in detail from social network analysis, behavior prediction analysis, and product recommendation applications.

Before the analysis, a representative graph database — Neo4j — should be introduced. Neo4j adequately stores nodes (target objects), node attributes, and node relationships. For example, Xiao Hong can be counted as a node, while her preference for red can be regarded as an attribute, and her being good friends with Xiao Hua is a relationship between nodes. Through the nodes, one can mine the relationship with each other. Therefore, even for complex relationships, one can achieve a fast query only by adding nodes.

In social network analysis, for a person's social relationship, Neo4j can establish a first-degree relationship (direct relationship), a second-degree relationship, etc., indicate the degree of intimacy, and finally establish a circle of friends (Figure 3.2). This is an important application of graph computing in social network analysis. According to the established social graph, not only can the relationship between the two of them be determined at a glance, but also friend suggestions can be made accordingly.

For behavior analysis, anti-fraud has become an important application of graph models. Specifically, fraudulent groups can be identified

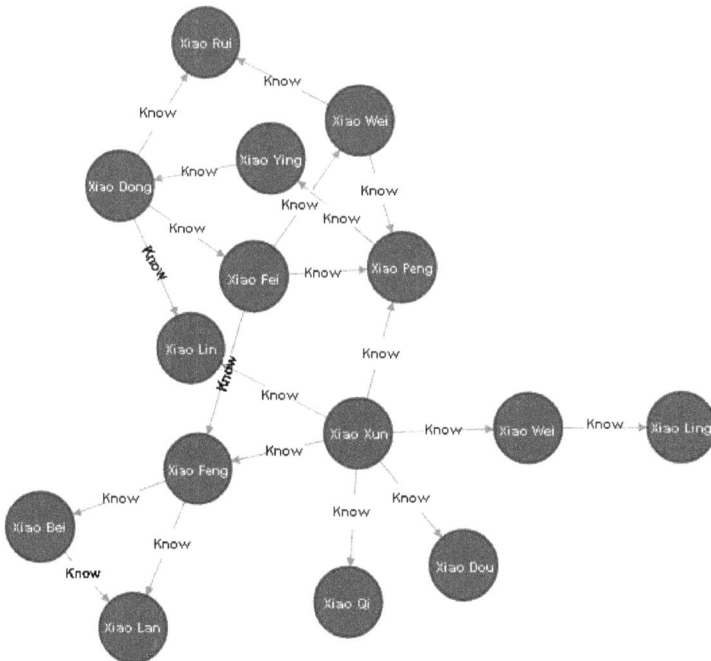

Figure 3.2   Building a Circle of Friends through Neo4j

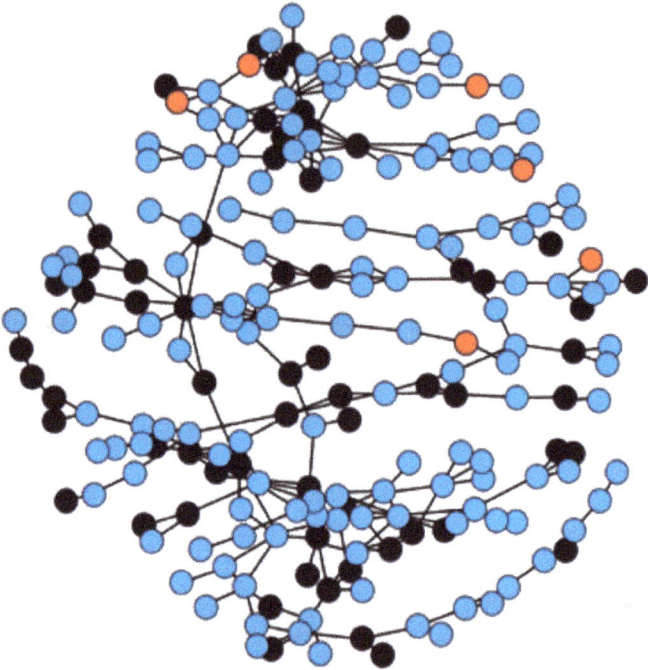

Figure 3.3   Fraud Team Graph

according to certain rules based on user characteristics. Figure 3.3 shows a typical fraud team graph. In the process of initiating investigation and verification, blue is the rejected user, black is the accepted but overdue user, and red is the passed and good user. It can be seen that blue occupies the vast majority, while the vast majority of those who pass have overdue behavior, which further verifies its fraudulent nature.

In addition, for the relationship between people and things, similarity recommendation is a typical example. When a user browses or buys a certain product on the Internet, we can establish a user's rating map for different items based on this, and predict and recommend similar high-scoring items with high similarity to the user. For example, if a target user buys a certain brand of shampoo online today, we can recommend the corresponding brand of toiletries. This is one of the applications of graph computing, and it is also a concrete manifestation of the idea of collaborative filtering in big data.

Stream model

Stream processing is like the operation process of a factory production line, which continuously supplies raw materials and finally determines the finished product through the chain of processing. In stream computing, the raw material is data, and each processing component can be regarded as a processing unit. The biggest advantage is the strong parallelization, while the biggest problem is that any problems in any chain will affect the subsequent work.

The basic idea of streaming is that the value of data decreases over time. Therefore, the stream processing mode always analyzes the latest data as quickly as possible and gives the analysis results, that is, real-time computing, as much as possible. Real-time calculation of data is a very challenging task. The data flow itself has the characteristics of continuous arrival, fast speed, and huge scale. Therefore, the data are usually not stored permanently. Moreover, since the data environment is constantly changing, it is difficult for the system to get an accurate picture of the entire data. Due to the requirement of response time, the process of stream processing is basically completed in memory, where memory capacity is a major bottleneck for stream processing.

The stream processing mode is suitable for scenarios that have strict real-time requirements, where the data do not need to be stored first and can be calculated directly. The requirements for data accuracy are often less stringent. The stream processing mode has many application scenarios, including the Internet and the Internet of Things, and is widely used in many fields. Here are three typical applications.

(1) *Real-time personalized recommendation system*: Real-time personalized recommendation systems can analyze the data generated by users in real time through streaming processing and make more accurate recommendations. At the same time, it can also provide feedback based on real-time recommendation results to improve the recommendation model and improve system performance. Real-time personalized recommendation systems are widely used. In addition to the e-commerce field that we are familiar with, it also includes the personalized recommendation of news, music, etc.

(2) *Business intelligence*: There are various data in the enterprise, such as inventory data, sales data, transaction data, customer data, and mobile terminal data. Business people usually want to efficiently

manage large amounts of data, obtain correct and complete information, and get answers in real time. Streaming is the best way to solve these problems. Through the calculation, the real-time data of each system within the enterprise can be obtained in real time, the monitoring and optimization of the global state can be realized, and the business decision support can be provided. The application of business intelligence is particularly extensive in transaction-intensive enterprises, such as banks, telecommunications companies, and securities companies.

(3) *Real-time monitoring*: Monitoring applications are generally associated with the Internet of Things. Various sensors transmit monitoring data in real time and at high speed. Through stream processing, monitoring data can be analyzed, mined, and displayed in real time. For example, in traffic monitoring, the traffic supervision department of each city generates massive amounts of video data every day, which are continuously input into the system in the form of streams. For another example, in the process of automatic operation and maintenance, the streaming system processes the operation and maintenance data in real time and generates an early warning.

Parallel decomposition model

Since some data are mostly semi-structured or unstructured, it is inconvenient to establish a strict relational model. Or even though a relational model has been established based on such data, the tasks need to be decomposed in parallel and processed by a huge number of machines because of the massive data. To address this, MapReduce undertakes the key role, decomposes the task into multiple sub-tasks, and processes them separately before summarizing.

Distributed processing is like the project leader assigning tasks to multiple departments, where department 1 does task A, department 2 does task B, and department 3 does task C. Multiple departments complete multiple tasks at the same time. Then, we summarize and reprocess the work results of each department, and finally get the result. MapReduce is a typical parallel decomposition model.

For example, if we need to find someone's specific location, we can search the database of each district separately and summarize the results. This is the most basic idea of parallel distributed processing, in which Map divides tasks into different machines for processing.

The task is divided into subtasks, and then Reduce will gather the processed results.

MapReduce is a parallel programming model for processing massive data and is used for parallel operations on large-scale datasets. MapReduce has the commonality of functional and vector programming languages, which makes this programming mode especially suitable for the search, mining, analysis, and machine intelligence learning of unstructured and structured massive data.

According to statistics, Google's backend server will perform 10^{11} operations each time its search engine is used. With such a huge amount of computing, if there is no good load balancing mechanism, the utilization rate of some servers will be very low, while some others will be overloaded and may even crash. These will affect the system's service quality. The use of MapReduce maintains the balance between servers and improves overall efficiency. Its core design ideas lie in two points: one is to divide and address the problem; the second is to effectively avoid a large amount of communication overhead during data transmission. Meanwhile, the MapReduce model is simple. In practical applications, many problems can be represented by the MapReduce model, such as building a search engine, web data mining, log data processing, and so on.

Typical Big Data Processing Systems

Data are ubiquitous and gradually become the basis for our decisions: We need data for shopping on Taobao; HR needs job applicant data to identify suitable candidates; we need to analyze geological information data to find oil at a lower cost. In this regard, some people may propose that what we want can be easily obtained through algorithm analysis. It is true that algorithms can help solve many problems, but with the explosive growth of data, it has become more and more difficult to find patterns through algorithms on a single machine. To this end, the distributed computing framework Hadoop came into being.

It is difficult to solve massive data demands on one machine. Hadoop, a distributed system infrastructure developed by the Apache Software Foundation, can distribute processing tasks to multiple machines for the same problem, perform task division and simultaneous processing, and finally get the answer collaboratively. It has many advantages, such as high efficiency and high fault tolerance.

Hadoop

How to understand Hadoop

There are countless keywords that everyone searches on Baidu every day. If one wants to know the number of searches for a certain keyword on Baidu in a certain period, how to get the results through Hadoop? First of all, it needs to be clear that Baidu cannot store all these keywords in memory but on multiple servers. For ease of explanation, we number these machines as follows: 1, 2, 3,..., n.

As shown in Figure 3.4, if we want to count the number of occurrences of the keywords, "lotus", "dictionary", and "beer" in machine 1, we can map, sort, and simplify the keywords entered such as "Phone, Calendar, Beer", "Dictionary, Lotus, Weather", and "Dictionaries, Rivers, Weather" during users' search. Finally, the number of each keyword is extracted (for convenience, it is assumed here that only three groups of keywords are searched).

The above is only the data on one machine, so how can we count the number of searches for each keyword on all machines? At this time, you can find another set of machines and call them a, b, c,..., n. We can let machine a count the keyword "lotus" that appears in all searches on the

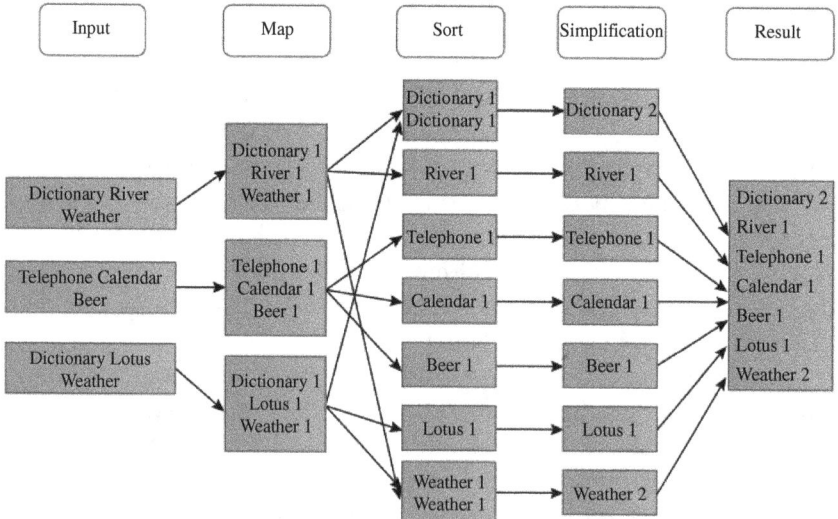

Figure 3.4   Statistical Keywords by Hadoop

machines, while machine *b* is responsible for counting the keyword "dictionary", and machine *c* is responsible for counting the keyword "beer". In a similar fashion, the number of occurrences of different keywords on all machines is eventually counted. In this process, machines *a*, *b*, *c*,…, *n* need to communicate and cooperate with each other. Neither several machines can search for a keyword at the same time nor can there be a situation where some keywords are not searched.

In January 2008, Hadoop became a top-level Apache project, and through this opportunity, Hadoop was successfully used by many companies other than Yahoo, such as Last.fm, Facebook, and *The New York Times*. The reason Hadoop can be widely used in big data processing applications is mainly due to its natural advantages in data extraction, transformation, and loading (ETL).

Hadoop architecture

The entire Hadoop ecosystem consists of many distributed components (Figure 3.5). There are 19 commonly used distributed components, making Hadoop supports more functions and constituting a big data processing system. However, Hadoop itself only includes the distributed file system HDFS, the distributed resource manager YARN, and the distributed parallel processing MapReduce.

For ease of understanding, we can do an analogy analysis. A Hadoop can be considered as supported by three stubs, HDFS, YARN, and MapReduce (Figure 3.6). Among them, HDFS, as the basis of the data

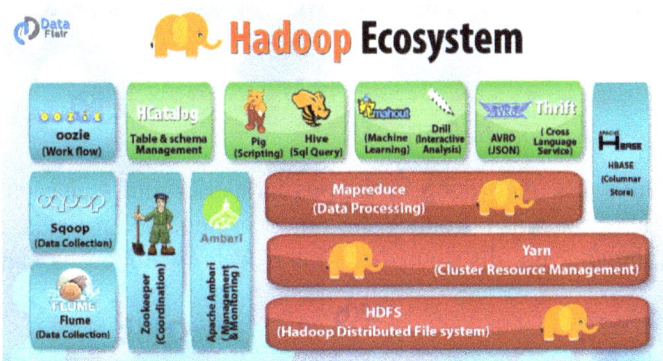

Figure 3.5   Hadoop Architecture Diagram

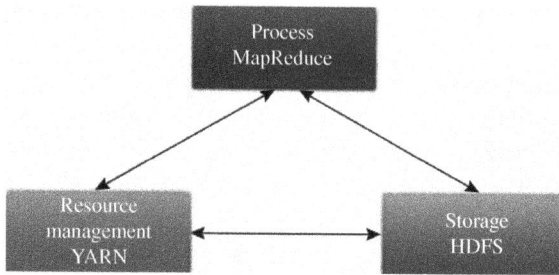

Figure 3.6   A Hadoop with Three Linked Roles

processing system, can be regarded as the data storage center of Hadoop, which is equivalent to the C drive and D drive of the computer. YARN is responsible for the management and scheduling of resources, which can be regarded as a computer with an operating system installed. Various applications can be run freely once it is installed. As a programming model, MapReduce defines data processing operations, which is equivalent to installing software development programs on the computer. This has greatly facilitated people to run programs on distributed systems without the need for distributed parallel programming.

From a functional point of view, Hadoop cannot do everything. It is actually a distributed infrastructure that provides the most basic and core functions in a distributed environment — distributed storage and distributed resource management. Obviously, the other components are utility components developed in the out layer of Hadoop. However, because the functions of the distributed database Hbase and the distributed lock service ZooKeeper are very important, we also classify these two components as the core components of Hadoop.

Hadoop application fields

The effectiveness of Hadoop in various industries and scenarios is well known. Here, we will introduce the practical application fields, such as building large-scale distributed clusters, data warehouses, and data mining.

(1) *Build a large-scale distributed cluster*: The most direct application of Hadoop is to build large-scale distributed clusters to provide massive

storage and computing services. For example, China Mobile's "Big Cloud" and Taobao's "Cloud Ladder" are already large or even super-large distributed clusters.

(2) *Data warehouse*: Many companies' log files and other semi-structured business data are not suitable to be stored in relational databases. However, these are especially suitable to be stored in semi-structured HDFS. Through applying other tools (such as data warehouse tool Hive and distributed database Hbase), one can provide services such as report queries.

(3) *Data mining*: In fact, data mining in the big data environment has not changed much. However, big data has caused problems for the pre-processing tools of data mining. Limited by the hard disk performance and the memory size, it takes at least 20 minutes for an ordinary server to read 1 TB of data, while Hadoop, on the other hand, increases the speed of reading and processing per server exponentially.

Spark

Now let's introduce Spark. The two frameworks, Hadoop and Spark, are often compared by industry insiders.

Why we need Spark

In data processing, we have discussed the distributed computing framework MapReduce. MapReduce divides tasks with Map and then summarizes through Reduce to output the results. Although it can achieve the expected results, MapReduce, as a batch engine, prefers to divide things into sections and process them in sequence. The results of each section need to be written to disk, which will be used in the following section again. As a result, the efficiency will inevitably be affected.

In this regard, MapReduce's "cousin", distributed in-memory computing framework, Spark, is completely superior to MapReduce. In 2009, the AMPLab at the University of California, Berkeley, developed the first Spark version. Since then, Spark has quickly occupied the big data processing framework market and has now become a top open-source project under the Apache Software Foundation. Its ecosystem is also becoming more and more perfect and has been in an absolute leading position in the field of data analysis.

Although both are for data processing, Spark is more flexible than MapReduce. It evaluates the tasks to be processed in advance, and then makes strategies and completes the tasks through the directed acyclic graph (DAG). In this process, there is no need to perform disk reading and writing, and various decomposition tasks can be done simultaneously or one by one, which thus greatly improves the speed. In terms of speed, Spark's batch processing is 10 times faster than MapReduce. However, since data are stored in memory, Spark has high requirements for memory.

Meanwhile, in terms of fault tolerance (which can be understood as the ability to quickly recover from errors), Spark uses Resilient Distributed Datasets (RDD), which are fault-tolerant collections, where if a part is lost or faulty, it can be quickly reconstructed. For MapReduce, it has to be recomputed. As for the security mechanism, MapReduce is superior to Spark since it integrates the security mechanism supported by Hadoop and other security projects of Hadoop.

In terms of programming languages, Spark supports development programs such as Scala, Java, Python, and R, allowing developers to work in their familiar language environment. At the same time, Spark applications are significantly smaller in code size than Hadoop, making them easier to program with less program development effort and better code readability and maintainability. On the contrary, MapReduce is not easy to program, but it can be applied with other tools.

In terms of compatibility (which can be simply understood as whether several objects can work together), Spark and MapReduce can not only work with YARN but also be compatible with other data, files, and tools through JDBC and ODC. Both have excellent compatibility. In terms of scaling, both can be scaled using HDFS. Spark has rich ecology, powerful functions, and a wide range of applications, while MapReduce is simpler and more stable.

In conclusion, Spark is an open-source, in-memory computing-based, fast, and general big data parallel computing framework running on distributed clusters; it greatly improves the real-time performance of data processing in the big data environment, ensuring high fault tolerance and high scalability; it also allows users to deploy Spark on a large number of cheap hardware clusters to provide cost-effective big data computing solutions.

Spark distributed computing process

Specifically, in a distributed situation, how does Spark gather resources and work together to solve problems? In this process, it is necessary to clarify where the tasks come from, how they are assigned, how they are executed, and how the results are feedbacked. As shown in Figure 3.7, the task mainly comes from the Driver, that is, the Driver puts forward the processing requirements first. It is equivalent to the case of advertisement planning, where Party A proposes the requirements first.

After the specific tasks are defined, the director of the advertising department will carry out task allocation and execution monitoring through YARN, including resource allocation and algorithm selection. The specific execution is completed by each Worker, who will allocate resources such as disk, memory, and network to process the task. This process is similar to the specific writing stage of the advertisement. The director of the advertising department divides the work, and the designers, copywriters, and other Workers undertake different tasks.

Just as in advertising planning, the design work will be further divided and handed over to different colleagues to complete the division of labor and collaboration. In the process of executing distributed tasks, Worker will also fully deploy personnel and establish *Executors* to further refine the tasks. The tasks are to be conducted in parallel, if possible, else they

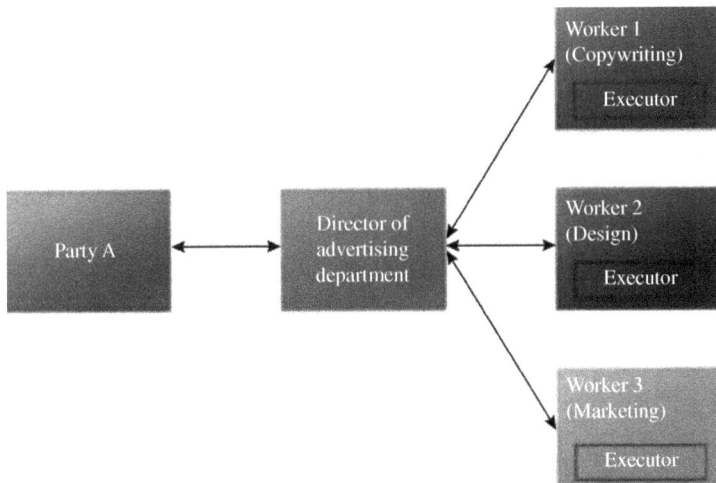

Figure 3.7 Spark Distributed Computing Flow

will be conducted in serial. At the same time, the execution result of Executor will be fed back to YARN and Driver by Worker to complete the whole process.

Spark application cases

With the growth of enterprise data volume, the processing and analysis of big data have become an urgent need. In this context, Spark has attracted widespread interest from academia and industry, and a large number of applications have landed in the industry. Many research institutions have started research works on Spark. The following will select representative Spark application cases for analysis so that readers can understand the application status of Spark in the industry.

Tencent Social Ads (formerly known as Guangdiantong) is one of the first applications that use Spark. Tencent's accurate recommendation based on big data takes advantage of the rapid iteration of Spark. It focused on the technical solution of "data + algorithm + system", and realized the whole process of "real-time data collection, real-time algorithm training, and real-time system prediction". The algorithm was finally successfully applied to the Guangdiantong pCTR delivery system, supporting tens of billions of requests per day.

Ali's search and advertising business initially used Mahout or MR written by itself to solve complex machine learning, which was inefficient and the code was not easy to maintain. Later, the Taobao technical team used Spark to solve multiple iterative machine learning algorithms, high computational complexity algorithms, etc. They applied Spark to Taobao's recommendation-related algorithms and also used Graphx to solve many production problems.

When Youku Tudou is using Hadoop clusters, the prominent problems include the following: (1) In terms of business intelligence, how to make user portraits faster, achieve precise marketing, and achieve accurate advertising recommendations? (2) Big data computation is very large and requires high efficiency. (3) Iterative operations of machine learning and graph computing require a lot of resources and are very slow. By comparison, it is then found that Spark is much superior to MapReduce: The interactive query response is fast, and the performance is several times higher than that of Hadoop; the simulated advertising has high calculation efficiency and low delay; the iterative computing such as machine learning and graph computing greatly reduces network transmission, data landing, etc., and greatly improves computing performance. At present, Spark has been widely used in Youku Tudou's video recommendation and advertising business.

Big Data Storage

As the "oil" of the "big data era", data are a massive asset that needs to be properly stored and kept. If there is a lack of reliable storage support, the data will become a tree without roots, and it is thus difficult to carry out subsequent mining and analysis. In data storage applications, various data storage methods and media selections are involved according to different scenarios.

Big data needs big storage

The data volume of big data is extremely large, which cannot be accommodated by a few terabyte-level hard disks. Moreover, its data growth rate has shown explosive growth. The traditional storage architecture can no longer meet the storage requirements of such a large amount of data. Therefore, it is necessary to understand how to meet the storage requirements of big data.

Before the emergence of cloud computing, the cost of data storage was very high. For example, to build a website, a company needs to purchase and deploy servers, arrange technicians to maintain the servers, and ensure the security of data storage and the smoothness of data transmission. They also need to regularly clean up data to free up space for storing new data. The overall labor and management costs of the computer room are high.

With the emergence of cloud computing, data storage services have derived new business models, and the emergence of data centers has reduced the company's computing and storage costs. For example, for a company to build a website, it doesn't need to buy a server or hire a technician to maintain the server. Instead, it can rent hardware equipment. The decline in storage costs has also changed people's view of data, and they are more willing to save historical data for one, two years, or even longer. With the precipitation of historical data, one could discover the correlation and value between data through comparison. It is possible to provide the best infrastructure for big data storage just because the emergence of cloud computing reduces the cost of data storage.

Petabyte-level big data storage

In the traditional data storage era, minicomputers are used for small data, and mainframes are used for big data. The storage size is proportional to machine size. However, today's explosive growth of massive data has far exceeded expectations. As a result, the traditional-scale servers for storage are lower than the demands.

The basic principle of cloud computing is to distribute computing on a large number of distributed computers instead of local computers or remote servers so that enterprises can switch resources to required applications and access computers and storage systems as required. When the cloud computing system centrally computes and processes big data, the cloud computing system needs to be configured with a large number of storage devices. Consequently, the cloud computing system is transformed into a storage system. This is how the concept of "cloud storage" emerged.

As early as 2008, Ren Yuxiang, chief architect of EMC's China R&D center, put forward the point of view that "the starting point of cloud storage should be petabytes". In 2012, China's cloud computing entered the first year of practice, with solutions and technical architectures emerging one after another. As the primary consideration for storage services, storage vendors have also pushed the capacity of cloud storage to a larger scale.

Traditional SAN or NAS have bottlenecks in capacity and performance expansion for the petabyte-level mass storage requirements. However, the emerging cluster storage based on cloud computing has the incomparable advantages of traditional network storage in terms of cost and performance, and has become a new favorite for customers pursuing high-cost performance. Compared with the use of dedicated servers, the A8000 ultra-low power cloud storage system launched by Yunchuang can carry a total storage capacity of up to 3.8 PB in a single cabinet, and the peak power consumption of a single storage node is less than 0.15 kW. It is 3 times more energy-efficient than traditional cloud storage products, and the construction cost of the storage system can be reduced by 5 to 10 times. The larger the storage scale is, the more obvious the advantages are and therefore could completely avoid the performance bottleneck of traditional storage.

Row storage and column storage

Today's data processing can be broadly divided into two categories: Online Transaction Processing (OLTP) and Online Analytical Processing (OLAP). OLTP is the main application of the traditional relational database, which is used to perform some basic and daily transaction processing, such as adding, deleting, changing, and checking database records. OLAP, on the other hand, is the main application of distributed database. It does not require high real-time performance but handles a large amount of data and therefore is usually applied to complex dynamic reporting systems. The difference between the two is shown in Table 3.1.

Why are there significant differences between OLTP and OLAP in the application categories of databases? In fact, this is due to the different database storage modes.

Traditional relational databases, such as Oracle, DB2, MySQL, and SQL Server, use row storage. In a row-based database, data are stored in logical storage units based on row data. The data in each row exist in the storage medium in contiguous storage, as shown in Figure 3.8.

Column storage is relative to row storage, and the emerging distributed databases such as Hbase, HP Vertica, and EMC Greenplum all use column storage. In a database based on column storage, data are stored according to a column as the base logical storage unit, and the data in a

Table 3.1   Differences between OLTP and OLAP

Contrast parameters	Comparative item A	Comparative item B
Type of data processing	OLTP	OLAP
The main object-oriented	Business developer	Analysis decision maker
Function realization	Deal with routine matters	xxx orientated analysis and decision
Data model	Relational model	Multidimensional model
The amount of data processed	Usually a few or dozens of records	Usually up to one or ten million records
Operation type	Query, insert, update, delete	Query-oriented

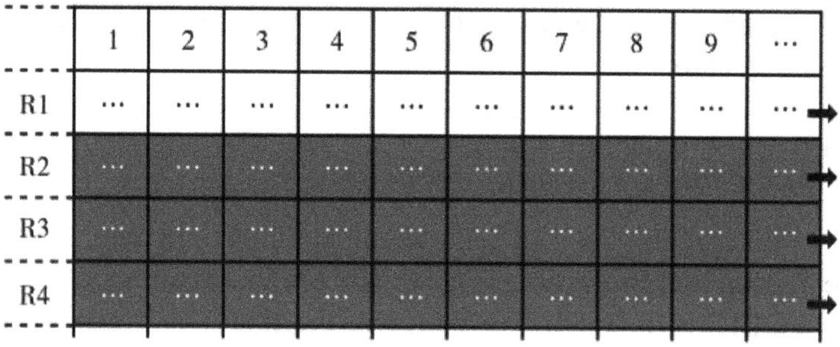

	1	2	3	4	5	6	7	8	9	...
R1	...	...	...	...	...	...	...	...	...	...
R2	...	...	...	...	...	...	...	...	...	...
R3	...	...	...	...	...	...	...	...	...	...
R4	...	...	...	...	...	...	...	...	...	...

Figure 3.8   Row Storage

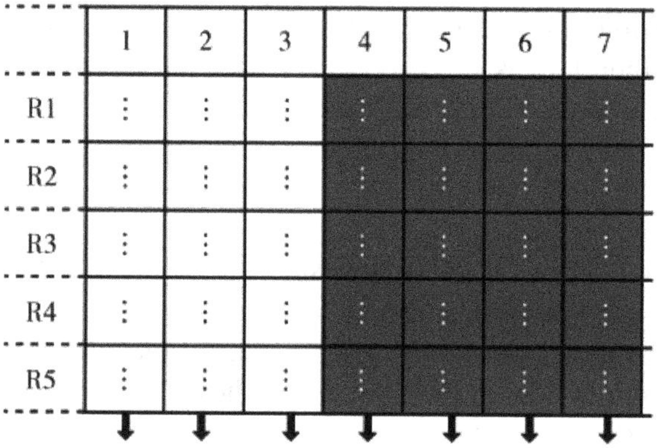

	1	2	3	4	5	6	7
R1	:	:	:	:	:	:	:
R2	:	:	:	:	:	:	:
R3	:	:	:	:	:	:	:
R4	:	:	:	:	:	:	:
R5	:	:	:	:	:	:	:

Figure 3.9   Column Storage

column exist as continuous storage in the storage medium, as shown in Figure 3.9.

The data of a table in the row storage are kept together, while those in the column storage are stored separately. Therefore, the data integrity in the row storage is very good, and it is easier to insert and update data. However, for row storage, all data will be searched even if only a few columns are involved in each query. In contrast, in column storage query, only the columns involved will be read. Any column can be used as an index. As a result, this can significantly reduce input/output (I/O) consumption and reduce query response time, yet the corresponding insert and update data are more troublesome.

Both row storage and column storage have their own advantages and disadvantages. For big data, if data integrity and reliability are the primary consideration, then row storage is the best choice. If data storage is the focus, the write performance of row storage is superior to that of column storage. For applications that require frequent reading of single-column collection data, column storage is more suitable. The applicable scenarios of row storage include transactions where data need to be updated frequently, small databases with few column attributes in the table, and/or real-time operations with deletes and updates. For column storage, its applicable scenarios include storage of massive message logs, user operation logs that can be accurately queried, and high requirements for query response time.

With the development of the column database, the traditional row database has added the support of columnar storage to the dual storage mode database system. For example, Oracle 12c launched the In Memory component, which enables the Oracle database for dual-mode data storage, thereby enabling the support of mixed-type applications. Of course, the columnar database has also added the support for row-based storage, such as HP Vertica. In short, there is no one-size-fits-all database storage model. The best storage mode always depends on the actual data storage and analysis needs.

Chapter 4

Big Data Management

Data management is the core of data processing. It refers to the various processes of collecting, sorting, organizing, storing, processing, transmitting, and retrieving different types of data.

Data management is an important application field of computers. It has two main purposes: (1) we can extract and derive valuable information from massive original data, and then use this as the basis for our own actions and decisions; (2) with the help of computer system, we can scientifically save and manage complex and massive data so that people can make full use of these information resources conveniently.

With the development of information technology, data management has gone through three stages: manual management stage, file management stage, and database management stage.

The History of Data Management

With the development of information technology, data management has experienced manual management stage, file management stage, and database management stage.

Manual management stage

Before the mid-to-late 1950s was the stage of manual management, which was the initial stage of computer management. The management of data was based on the programmer's personal considerations and

| Application program 1 | ————————▶ | Data 1 |

| Application program 2 | ————————▶ | Data 2 |

⋮   ⋮

| Application program *n* | ————————▶ | Data *n* |

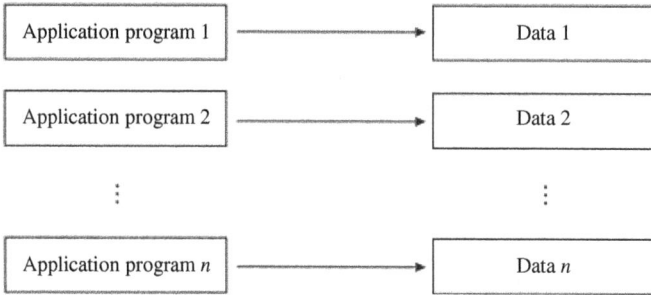

Figure 4.1   The Relationship between Programs and Data

subjective ideas. Programmers should consider data structures, storage addresses, storage methods, and input and output formats when compiling programs. If the storage location of data or the format of input and output changes, the corresponding program will also change accordingly. When people use the system for data processing, they need to prepare data every time. The characteristics of this stage are the following: the data and the program are closely integrated into a whole, a set of data corresponds to a program, and the data are not independent. The relationship between programs and data is shown in Figure 4.1.

File management stage

From the late 1950s to the early 1970s, it was the stage of file management when the operating system emerged, which had made great development and progress compared with the previous stage. The operating system contains a special software for managing data, the file system. The file system organizes the data into an effective dataset according to certain rules, called data files. At this stage, data can be stored on the external storage device for a long time in the form of files, and the operating system such as data storage is automatically managed by the file system. The file management system becomes the interface between the application program and the data file. The relationship is shown in Figure 4.2.

Database management stage

After the late 1960s, it was the database management stage. At this stage, computer technology has been greatly developed, and there are

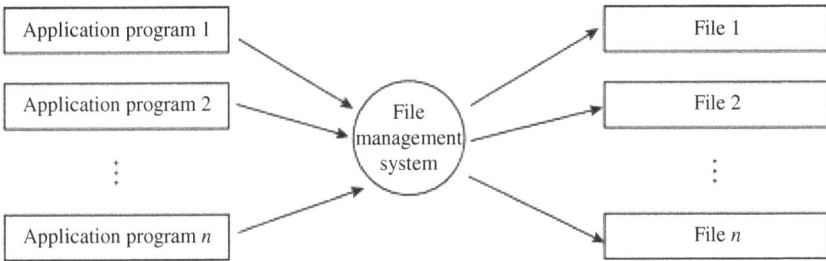

Figure 4.2   File Management System

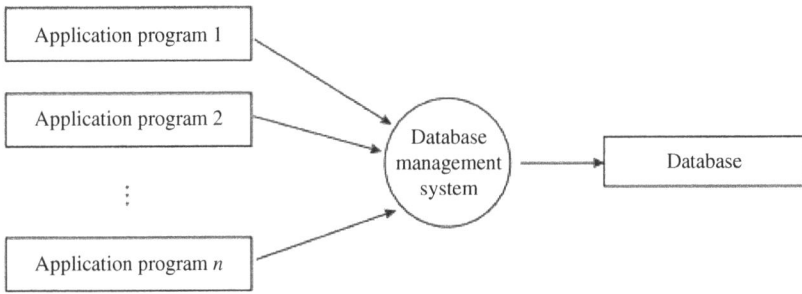

Figure 4.3   Database Management System

large-capacity disks in hardware. As computers are used in enterprise management, the amount of data increases rapidly, and it is urgent to centrally control data and provide data sharing. Therefore, a new data management method, namely database management software, has been developed. This technology avoids the shortcomings of the previous management methods and makes the database management technology enter a new stage. At this stage, the data management system becomes the interface between users and data, and its relationship is shown in Figure 4.3.

Big Data Management Approach

Previously, the data management method based on manual checking and processing on a table-by-table and item-by-item basis could no longer meet the demand for data value mining in the future due to heavy workload, poor quality, and low efficiency.

With the development of information technology, in recent years, information management has become an important direction of enterprise management, and more and more enterprises have begun to use information technology for data management. Through the big data management method, the informatization coverage of the whole business and all personnel can be realized. Specifically, there are the following eight main management methods.

Big data backup

Today, big data management and storage are breaking away from the physical world and rapidly entering the digital realm. With the development and progress of science and technology, the amount of big data has grown rapidly. At this rate, all the machines and warehouses in the world will not be able to fully accommodate it.

The adoption of cloud computing is also prospering as cloud storage drives digital transformation. Data are no longer risk-controlled in one location and can be accessed anytime, anywhere, and many large cloud computing companies will have more access to basic statistics. If data are backed up on these servers, years of data business growth and development will not be eliminated in the event of a network attack, and the cloud will provide unique services by migrating from A to B.

There are six types of big data backup: local backup, off-site backup, live backup, death backup, dynamic backup, and static backup. Among them, the most typical off-site backup is the "two locations and three centers", which is the most used backup method in banks.

Big data update

Data are continuously generated, collected, and loaded into big data analytics systems. Data analysis operations designed and optimized on static data, on the one hand, cannot reflect the latest data and are not suitable for the needs of many online applications; on the other hand, they may be interfered with by data update operations and cannot achieve optimal performance. Therefore, in the design of the big data analysis system, we must not only focus on the big data analysis operation itself but also take the big data from update to analysis as an important reference factor.

In order to support big data updates, the most basic requirement is to be able to store newly arrived and newly generated data. But this is far from enough. The more important thing is to effectively update the data to ensure the improvement of the efficiency of big data analysis.

Big data authorization

The authorization of data science refers to giving users the right to legally use scientific data. Whether it is free public sharing or conditional sharing, or data classification in a commercial environment, scientific data authorization is required.

Data science resources from multiple sources require diverse authorization methods. There are mainly four authorization modes: (1) Completely free sharing — these data belong to basic process data. (2) Conditional free sharing — the source of these data is not as wide as the data in the completely free sharing model, which belongs to public welfare or data resources with public attributes. (3) Recovery of service cost sharing — these data constitute a data resource unique to some business departments. Although it is generally available for free sharing, a fee will be charged if data service activities are carried. (4) Paid sharing — commercial operation is adopted, and data resources are used to make profits.

Big data index

Index is an important concept in relational database, which is a decentralized storage structure created to speed up the retrieval of data rows in a table. Index is built against table and consists of index pages other than data pages. The rows in each index page contain logical pointers to speed up the retrieval of physical data.

The index design in big data management mainly considers high performance and high scalability and can effectively support different types of queries. The main index structures include secondary indexes, double-level indexes, indexes sorted by spatial objects, and so on.

The establishment of an index in a database system has four main functions: (1) to quickly read data, (2) to ensure the uniqueness of data records, (3) to achieve referential integrity between tables, and (4) when using ORDER by and group by clauses for data retrieval, the use of indexes can reduce the time of sorting and grouping.

Data normalization

Data normalization refers to scaling data to a small, specific interval. It is often used in some comparison and evaluation index processing to remove the unit limitation of the data and convert it into a dimensionless pure value so that indicators of different units or magnitudes can be compared and weighted.

Before performing data analysis, we usually need to normalize the data first, and then use the normalized data to perform data analysis. Data normalization processing mainly includes two aspects: data homogenization and dimensionless processing. Data homogenization processing mainly solves data problems of different natures, while data dimensionless processing mainly solves the problem of data comparability. The methods of data normalization mainly include min–max normalization, z-score normalization, decimal scaling, and so on.

Data engineering

Data engineering is the application and destination of data science and data technology. It solves data problems in the real world with innovative ideas and uses engineering viewpoints for data management and analysis, as well as the development and application of systems, including data system design, data application, and data services.

The theoretical foundations of data engineering come from several different disciplines, including statistics, information science, computer science, artificial intelligence, information systems, and so on. Therefore, data engineering can support the research and application of big data. The purpose of data engineering is to systematically and deeply explore various scientific problems, technical problems, and engineering realization problems encountered in the application of big data, including data life cycle management, data management, analytical computing and algorithms, data system infrastructure construction, and big data application facilities and promotion.

Data cleaning

Data cleaning refers to the final process of "washing out dirty data" and finding and correcting identifiable errors in data files, including checking data consistency and handling invalid and missing values.

The data in a data warehouse constitute a collection of data on a certain subject, which comes from multiple business systems and sometimes even includes some historical data. As a result, some data are inevitably wrong, while some conflict with each other and more likely contain incomplete or duplicate data. Such problematic data are useless. Data cleaning aims to filter the "dirty" data that do not meet the requirements and submit the filtered results to the business department, which will confirm whether it needs to be filtered.

The purpose of data cleaning is to correct errors, normalize data formats, and remove abnormal and duplicate data by filling in missing data values, smoothing noise, identifying, or removing outliers, correcting data inconsistencies, and other methods.

Data maintenance

In the process of IT operation and maintenance, data operation and maintenance are very important. To maintain a stable operation of the data center, professional technicians at a senior professional level are necessary. The data center for critical business has technicians on duty 24 hours a day, while the unattended data center can generally only undertake unimportant business. There are almost no data centers that are completely not attended or maintained. The daily maintenance of the data center is cumbersome but necessary. At the same time, because data play an important role in daily work and life, the data center that carries the data calculation and operation is becoming more and more important, which also highlights the importance of maintenance work.

The maintenance work of a data center is generally divided into four categories: (1) daily inspection, (2) application change and deployment, (3) software and hardware upgrade, and (4) sudden failure handling. In IT infrastructure construction, once the data center is put into production, maintenance work is in place, until the end of the data center lifecycle.

Data Integration

For data mining, data are very important because users always hope to obtain as much target data as possible for mining. There's a "data integration" problem involved here. What is data integration?

Data integration is the integration of data from several scattered data sources, either logically or physically, into a unified dataset. These data sources include general files, data warehouses, and relational databases. The core task of data integration is to integrate interrelated distributed and heterogeneous data sources so that users can access these data sources in a transparent manner. Integration refers to maintaining the overall data consistency of the data source and improving the efficiency of information sharing and utilization. The transparent way means that users do not need to worry about how to access data from heterogeneous data sources but only focus on what kind of data is accessed in what way.

Issues involved in data integration

Data integration has three key problems that need to be solved, mainly including data value conflict, attribute redundancy, and entity recognition.

The first is the conflicting data value problem. Different representation, encoding, and specification units of attribute values will cause the same entity in the real world to have different attribute values in different data sources; same attribute names but with different meanings. Differences in semantics and data structures between attributes from different data sources bring great difficulties to data integration. Care needs to be taken to avoid redundancies and inconsistencies in the integrated dataset.

The second is the attribute redundancy problem. An attribute may be redundant if it can be derived from other attributes or a combination of them. To solve the redundancy problem, it is necessary to detect the correlation between attributes. For numerical attributes, the correlation between the two attributes is evaluated by calculating the correlation coefficient between them. For discrete data, we can use the chi-square test to do a similar calculation and judge whether the two attributes are independent assumptions based on the calculated confidence level. The redundancy of the tuple itself also constitutes data redundancy. Database designers sometimes use lower-level paradigm requirements for certain performance requirements, causing the same tuple in different relational tables to be stored in duplicate instead of using foreign key associations. In this way, there is still a risk of omission of updates, resulting in inconsistent content of copies.

The third is the entity recognition problem. The same entities from different data sources may have completely different names, so how can they be identified correctly? For example, cust_number and customer_id are from different databases, but the meanings conveyed are the same. Here, we can analyze them using the metadata of attributes. The metadata of an attribute generally includes the attribute name, meaning, data type, value range, data value code, and missing value symbol. Using attribute metadata for data cleaning could avoid schema integration errors.

Data integration application

cData data integration middleware accesses heterogeneous databases, legacy systems, Web resources, etc. through a unified global data model. The middleware is located between the heterogeneous data source system (data layer) and the application program (application layer). It backward-coordinates the data source systems and upward-provides a unified data schema and a generic interface for data access for applications accessing integrated data. The applications of various data sources still complete their tasks, and the middleware system mainly provides a high-level retrieval and integration service for heterogeneous data sources.

cData data integration middleware is a popular data integration method. It hides the underlying data details by providing a unified data logic view in the middle layer. Users can view the integrated data source as a unit. The key issue under this model is how to construct this logical view and make it possible to map different data sources to this middle layer.

Big Data Privacy Management

The combination of data analysis and modern psychology has had a great impact on society, especially in the field of the Internet. Psychological analysis based on big data can accurately judge people's personalities and psychological characteristics and other privacy.

Big data privacy applications

Data storage and processing costs continue to decline, technologies in data analysis and machine learning and other fields develop continually, and the application of data analysis is more and more extensive, prompting people to enter the era of big data. In this era, Internet giants collect a great amount of user information and use data analysis and algorithms to continuously optimize the displayed information, thereby influencing users' browsing and purchasing behavior. Through these methods, the advertisement placement of e-commerce platforms such as Tmall, JD.com, and Suning has become more and more accurate and effective, and their platform revenue is also increasing. Therefore, the issue of big data privacy has received increasing attention.

In addition, human behavior can be influenced or altered by subtle manipulations. For example, the study found that when other conditions remain unchanged, if the e-commerce company selling mattresses uses white clouds as the background of the web page, more customers will choose comfortable but higher-priced products; using coins as a backdrop would have the exact opposite result, leading more customers to choose less expensive products. Moreover, studies have found that addictive behaviors of humans can be triggered and reinforced by a variety of means. For example, positive feedback can be used to reinforce behavior, and if the occurrence of feedback is uncertain but maintained with a certain probability, it will have a greater reinforcement effect than definite feedback. Other reinforcement methods include creating suspense, adding social elements to feedback, and setting goals and their corresponding progress countermeasures. These methods are increasingly applied to the design of online products.

As data collection and analysis become increasingly important, the network and data have become one of the new battlegrounds for countries to compete for. For example, the United States and some European countries have accused Russia of using the Internet to conduct aggressive behavior against them, arguing that Russia is trying to influence their own political system. Therefore, network and information security has been regarded as an important part of national security. At the same time, laws lag behind innovations in technology and business models, and countries and regions such as the United States, China, and the European Union have different attitudes and perceptions of privacy, data protection, and data security.

Due to the requirements of information system security and confidentiality, networks of different security levels cannot be directly connected, resulting in the isolation of intranets, private networks, and public networks. The connection of intranet, private network, and public network may cause immeasurable losses once information leakage happens.

In order to solve the problem of cross-network communication security, Yunchuang Big Data independently developed the physical isolation one-way shutter, which not only plays the role of logical isolation but also of physical isolation. The physical isolation one-way shutter is based on the one-way transmission characteristics of physical light, and realizes physical isolation, one-way transmission, transparent and visible in the middle, without any reverse channel. It can ensure communication security and realize one-way reliable communication between the industry intranet, private network, and public network. What makes this technology so special is that it is transparent and visible in the middle part, with only one light-emitting end and one light-sensing end to ensure that there is no reverse channel, which is different from the traditional logic isolation device.

Personal data protection system

In the information age, the scale of data collection and data sharing has grown rapidly, and data has become one of the sources of economic growth and social value creation. The rapid development of data mining and utilization technology makes individuals gradually change from "anonymity" to "transparency" in cyberspace, making it difficult for traditional methods to effectively deal with the new problems of personal data protection in the big data environment. Reasonable and effective new institutional arrangements are a good means to solve the above problems. Therefore, following the international advantages of personal data protection and improving the personal information protection system and credit reporting system are not only the basis for using data mining to promote economic development but also the basis for improving the efficiency of financial services, expanding service coverage, and improving the service quality.

Personal data and public data are a set of relative concepts. In the contemporary context, it is necessary to clarify their definitions and attributes. Public data refers to data resources that are not identifiable, that is, not directed by the subject, such as the trend map of the flow of commodities and materials, and the map of the flow of social funds. In contrast, the identifiability of personal data constitutes an internationally recognized general characteristic of personal information.

Personal data generally refers to any information relating to an identified or potentially identified natural person, including personal name, address, date of birth, medical records, personnel records, ID number, and other specific personal information that can be identified alone or in comparison with other information.

Personal data has dual rights attributes. In terms of the legal rights attributes, personal data has characteristics similar to personality rights, such as name rights, portrait rights, and reputation rights. As society develops, personal data has gradually acquired the attributes of property rights, such as ownership, usufruct, and disposal. Therefore, it has become an international trend for personal data rights to be endowed with independent civil rights with dual attributes of personality rights and property rights.

At present, just like production factors such as land, capital, and labor, data resources are becoming the basic elements to promote global economic growth and social development of various countries. Objectively speaking, the mining and utilization of personal data by informatization is a double-edged sword. On the one hand, information technology is widely used, and the development and utilization of personal information are of great significance to the progress of social development. Commercial organizations can use the collected personal information to provide auxiliary information for production and marketing decisions. Government departments can also use the personal information they have to make accurate decisions, thereby improving the effectiveness of social management and preventing and punishing crimes. On the other hand, the improper collection, misuse, and leakage of personal information will cause problems, such as the gradual loss of data autonomy, violations of privacy, and unfairly distributed economic benefits of data. Therefore, the development and utilization of personal information on the legal track is the foundation of the fair development of the market economy.

In order to maximize the rational use of data on the basis of protecting personal data, China can learn from international trends and consider the actual national conditions to build a personal data protection system, and make systematic arrangements and plans in terms of ensuring legislation, effective supervision, and international cooperation, and improving the data operation protection system.

Goals of big data privacy management

The overall goal of big data privacy management is to use our own management concepts and methods to manage big data privacy like web data,

XML data, and mobile data. Specifically, we can start from the following three aspects:

(1) *Give individuals and businesses "reassurance"*: For those who want to disclose and plan to share data, data privacy comes first. Data can be made public or allowed to be accessed by other users without exposing data privacy, for example, disclosing personal social network information but avoiding the risk of losing a job and disclosing personal location for scientific research but avoiding the risk of malicious tracking.

(2) *Find ways for unresolved privacy challenges*: At present, many fields have not found suitable privacy protection strategies yet. The challenges include, for example, in the marketing field, how to ensure that consumers' information is not abused when making insurance decisions; in medical insurance and research fields, how to mine personal clinical data while avoiding the risk of insurance discrimination and how to distribute humanized genetic medicines and avoid the misuse of medical data.

(3) *Provide technical support for the application of big data*: Privacy is a prerequisite for the application of big data. If the privacy issue cannot be well resolved, the corresponding application is likely to become empty talk. For example, we need to prevent data collectors, analysts, and users of analysis results from maliciously leaking private information; also, we need to avoid privacy leakage in all stages of collection, processing, storage, conversion, and destruction in the big data life cycle.

In conclusion, the law and practice in the data field in China are influenced by many factors. On the one hand, the government's need for social stability largely takes precedence over privacy and data protection. For example, the government will clean up related vulgar content. On the other hand, the awareness of the public and users to safeguard their own interests has been continuously strengthened, which has promoted the continuous improvement of user privacy and data protection systems.

Chapter 5

Big Data Analysis Methods

In the face of an ever-growing flood of data, how to find valuable information in these huge, diverse, and complex data and apply these data reasonably and efficiently poses a challenge to the application of data analysis. In this process, mastering big data analysis methods and data visualization applications, so that big data can better serve our work and life, has gradually become the consensus of people.

Data Analysis Process

Through data analysis, Google can predict the trend of influenza outbreaks in a certain area and remind people to prepare for prevention in time; Taobao recommends personalized products for users based on their browsing and consumption record data; NetEase Music customizes playlists based on users' music records through similar algorithms. These applications that are close to life are all closely related to data analysis. Now, what does a complete data analysis process look like?

Problem definition

In the data analysis process, the first thing is the problem definition, that is, what problem is the follow-up data analysis work focused on. It includes specific issues to be analyzed, key influencing factors and core indicators, etc., and clearly defines the factors that need to be focused on and analyzed. The precise definition of the problem can greatly improve the efficiency of data analysis.

The problem definition process has high requirements for analysts' professional knowledge, data sensitivity, and data thinking. When we are new to a large sample of data, it can be overwhelming. As more and more problems are analyzed, analysts are gradually becoming more sensitive to data, gradually developing the habit of using data to analyze and speak with data. Based on data and experience, preliminary judgments and predictions can be made.

Data collection

Even if a data analyst is experienced and has a clear and rigorous data mindset, just like cooking a meal without rice, it is difficult to do nothing without the data to build on. Therefore, after identifying the problem, the first task is to obtain the required data by various means. If the identified problem is to conduct air quality analysis, then weather data, pollutant gas emission data, air quality index, etc. are all important references; if the identified problem is to analyze product sales performance, the historical sales data, user portrait data, advertising data, etc. need to be emphasized in analysis.

As for the specific data acquisition method, we can either use the internal database to complete database management such as data extraction with SQL skills, or obtain the open data sets from government, scientific research institutions, and enterprises. In addition, we can also write web crawlers to collect various data on the Internet such as Zhihu, Douban, and NetEase, and analyze a certain industry or a certain group of people, providing an important reference for market research and competitive product analysis.

Data preprocessing

After obtaining the required data, not every piece of data can be used directly. Many data are incomplete, inconsistent, and noisy "dirty" data, which requires data preprocessing before they can be used in the analysis and research. For example, when analyzing air quality, due to equipment reasons, some data may be repeatedly recorded, and some data may not be accurately recorded. Therefore, such data must be preprocessed first for accurate analysis.

During preprocessing, various methods such as data cleaning, data integration, data transformation, and data reduction can be used to process the data to good form to correct the inconsistency and incompleteness, transform or unify the data into a form suitable for mining, and finally obtain more precise analysis results.

After data preprocessing, data analysis and modeling need to be considered. At this stage, we should understand the applicable scenarios of different methods according to the defined problems and the determined data, that is, what kind of problems the methods can solve, the premise of the methods' application, the data requirements of the methods, etc., so as to choose the appropriate data analysis method, data mining algorithm, and optimize the analysis model. The abuse and misuse of statistical analysis methods can only be avoided by an accurate match between questions, data, and analysis methods.

It is worth mentioning that in data preprocessing, it is particularly important to choose the correct statistical analysis methods to conduct exploratory and repeated analysis. Each statistical analysis method has its own characteristics and limitations. It is unscientific to draw conclusions based on the results of only one analysis method. Therefore, in the process of analysis, it is often necessary to select several statistical analysis methods at the same time to repeatedly confirm the analysis.

Data visualization and report writing

Through data analysis, the final analysis result will be obtained. At this time, it will be accurately described and predicted, and the results will be displayed through data visualization, which becomes the final link of data analysis.

In a specific data analysis report, the analysis results need to be presented directly so that readers can establish a comprehensive understanding of the relevant situation. Specifically, in order to provide convincing results in the analysis report, a storytelling logic should be established, and the analysis should be conducted in depth from different angles and aspects, and also going deep and detailing all aspects.

Key Data Analysis Methods

In the process of data mining analysis, data analysis methods can be classified differently according to various standards. For example, the analysis methods often used in Internet operations include subdivision analysis, comparative analysis, funnel analysis, cohort analysis, cluster analysis, AB testing, buried point analysis, source analysis, user analysis, and form analysis. In enterprise data analysis, it can be specifically divided into descriptive analysis, diagnostic analysis, predictive analysis, and instructional analysis. If it is classified according to the analysis chart, it can be divided into various types such as histogram analysis, boxplot analysis, time series chart analysis, scatter plot analysis, comparison chart analysis, and funnel chart analysis.

Data mining is a multidisciplinary field involving database technology, artificial intelligence, high-performance computing, machine learning, pattern recognition, knowledge base engineering, neural networks, mathematical statistics, information retrieval, information visualization, and many other fields. It integrates knowledge and technology from many disciplines. Therefore, the current data mining methods and algorithms have presented an extremely rich form. In a broad sense of usage, the commonly used analysis methods in data mining mainly include classification, valuation, clustering, prediction, visualization, and association rules. Here, three methods of classification analysis, cluster analysis, and association analysis are briefly introduced.

Classification analysis

Classification is one of the key methods in data mining. Classification is to find out the common characteristics of data objects and divide them into different classes according to the classification mode. In other words, classification is a process of differentiating the labels of other sample data based on the training sample data (sample data that is pre-labeled), which can be understood as to label the sample data. One example is that car retailer organizes and analyzes existing customer data and divides it into different categories according to customers' preferences for cars. When customers with the same car preferences consult, the salesperson can introduce and analyze cars for specific customers in a targeted manner. It can greatly increase the transaction rate while reducing communication costs.

The difference between clustering and classification is that the classes divided by clustering are unknown, and there is no need to manually label or train the classifier in advance; for classification, the class needs to be defined in advance, and the sample data is input to construct the classifier.

Commonly used classification methods include decision tree, Bayes, artificial neural network, *K*-nearest neighbor, support vector machine, logistic regression, and random forest. The decision tree, Bayes, and artificial neural network methods are briefly introduced in the following.

Decision tree is an instance-based induction algorithm and is one of the main techniques for classification and prediction. It focuses on deriving classification rules represented by decision trees from a set of unordered and irregular instances. The purpose of constructing a decision tree is to find out the relationship between attributes and categories. The decision-making process of the decision tree needs to start from the root node and compare the data to be tested with the feature nodes one by one. Then, select the next comparison branch according to the comparison result and repeat until the leaf node makes the final decision. Figure 5.1 shows the decision process of a decision tree.

Bayesian classification algorithm is a class of algorithms that use probability and statistics for classification. For instance, naive Bayesian classification and other algorithms mainly use Bayes' theorem to predict the possibility of an unknown class of samples belonging to each class. The most likely category is selected as the final category of the sample. This type of algorithm can be generally used in large databases. Among them, naive Bayesian classification is conceptually simple. It is simple to use, yet very useful. Its classification logic can be understood as follows: for a given item to be classified, according to different preconditions, the probability of each category appearing under different circumstances is calculated, and whichever has the highest probability can be classified under the category. To give a simple example, when a dark-skinned

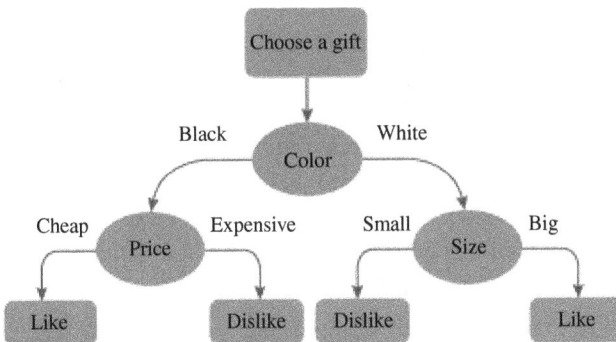

Figure 5.1   The Decision Process of a Decision Tree

person is approaching, most people will assume that he is African because dark-skinned people are most likely from Africa without further specific information.

Artificial neural network is a mathematical model for information processing by imitating the characteristics of biological neural network and applying a structure similar to the synaptic connection of brain. In artificial neural network, a large number of nodes (called "neurons" or "units") are connected to each other to form a "neural network" to complete the information processing. Humans have been studying artificial neural networks for long, such as the perceptron model in the 1950s and the convolutional neural network in the 1990s. For neural networks, training of the network is usually required. In this process, the training changes the value of the connection weight of the network nodes, making it a function of classification. Consequently, the trained network can be used for object identification. At present, although there are hundreds of common neural networks such as BP network, Hopfield network, radial basis RBF network, random neural network, and competitive neural network, there are still common shortcomings, such as slow convergence speed, long training time, large amount of computation, and inexplicability.

Cluster analysis

With the development of technology, data collection has become relatively easy, thus resulting in increasingly larger databases, for example, various online transaction data, and image and video data. The data can usually reach hundreds or even thousands of dimensions. In natural society, there exist a lot of data clustering problems.

Clustering refers to the process of grouping physical or abstract objects into multiple classes through collections, that is, classifying data into different classes or clusters. Objects in the same cluster have great similarity, while those in different clusters have great dissimilarity.

Clustering is an active research field and is one of the main tasks of data mining. Clustering can be used as an independent tool to obtain the distribution of data, observe the characteristics of each data cluster, focus on further analysis of a specific cluster set, and can also be used as a preprocessing link for other algorithms.

Traditional clustering algorithms can be divided into five categories: partitioning methods, hierarchical methods, density-based methods, grid-based methods, and model-based methods. The abovementioned algorithms have relatively successfully solved the low-dimensional data clustering problem. However, many existing algorithms have failed for high-dimensional data and large data.

Correlation analysis

Association analysis is a simple, practical technique that aims to discover associations or correlations that exist in large data sets, thereby describing the regularity and pattern of simultaneous occurrence of certain attributes in a transaction. In the field of data mining, association analysis is called association rule mining.

Association analysis is to discover associations and correlations between item sets from a large amount of data. A typical application is shopping basket analysis, which analyzes customers' purchasing habits by discovering the connections between different items that customers put in shopping baskets. Among them, understanding the products that are frequently purchased by customers at the same time could help retailers to further develop their marketing strategies. For example, the bundling of beer and diapers was discovered and established. In addition, price list design, product promotion, product placement, and customer segmentation are also important applications.

The algorithms of association analysis mainly include breadth-first algorithm and depth-first algorithm. Among them, breadth-first algorithms include Apriori algorithm, AprioriTid algorithm, AprioriHybrid algorithm, Partition algorithm, Sampling algorithm, and Dynamic Itemset Counting (DIC) algorithm. The depth-first algorithm mainly includes FP-growth algorithm, Equivalence Class Transformation (Eclat) algorithm, and H-Mine algorithm.

Big Data Visualization

After clarifying the process and specific methods of data analysis, further data visualization analysis is required to display the results of data analysis clearly and accurately. In the analysis process, the user is the subject of all behaviors: visual information is acquired through the visual perception organ, and then is encoded and formed into cognition. Different data visualization methods have different intuitive effects. According to different principles, there are various classifications of data visualization methods.

For example, according to different fields, it can be divided into geographic visualization, life science visualization, network and system security visualization, financial visualization, etc.; according to various spatial dimensions, it can be divided into one-dimensional visualization, two-dimensional visualization, three-dimensional visualization, complex high-dimensional visualization, etc.; according to different visualization objects, it can be divided into text and document visualization, cross-media visualization, hierarchy and network visualization, etc. From a methodological point of view, data visualization methods can be divided into three levels, as shown in Figure 5.2.

Statistical chart visualization method

Statistical charts are not only the earliest form of data visualization, but also a basic visualization technique. Diverse charts such as column charts, line charts, pie charts, area charts, maps, word clouds, waterfall charts, and funnel charts have been widely used. For many complex large-scale visualization systems, such charts have become an indispensable basic element. Choosing the appropriate combination of statistical charts and visual cues will help the realization of data visualization and meet different display and analysis needs. Figure 5.3 summarizes the statistical visualization methods available.

The most important purpose and highest pursuit of data visualization are to represent complex data relationships in a simple and easy-to-understand visualization form. Basic visualization charts can meet the needs of most visualization projects. In the following, we will introduce the

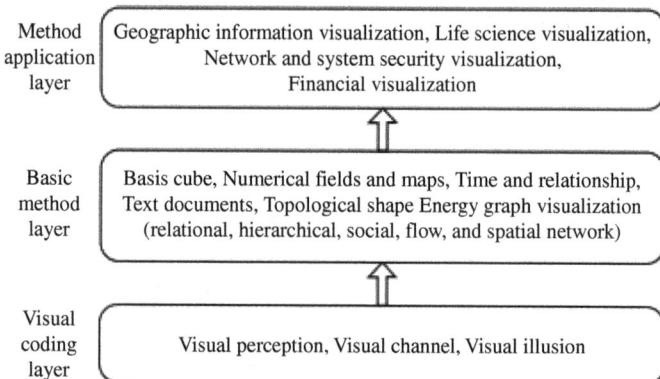

Method application layer	Geographic information visualization, Life science visualization, Network and system security visualization, Financial visualization
Basic method layer	Basis cube, Numerical fields and maps, Time and relationship, Text documents, Topological shape Energy graph visualization (relational, hierarchical, social, flow, and spatial network)
Visual coding layer	Visual perception, Visual channel, Visual illusion

Figure 5.2   Data Visualization Method System

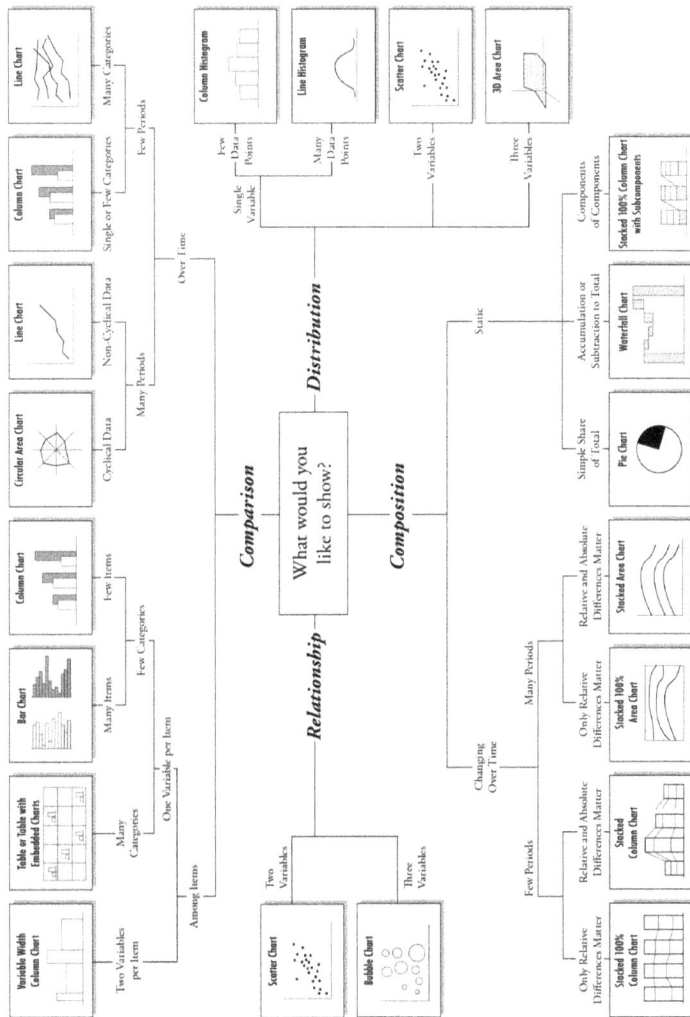

Chart Suggestions—A Thought-Starter

Figure 5.3 Statistical Visualization Method

commonly used visual charts — column charts, bar charts, line charts, pie charts, scatter charts, and radar charts.

Column chart

A column chart is a statistical report graph that expresses a graph with the length of a rectangle as a variable. It represents the distribution of data by a series of vertical stripes with unequal heights to compare two or more values.

Column charts have only one variable and are suitable for two-dimensional datasets to clearly compare data in two dimensions. It can also be arranged horizontally or expressed in a multi-dimensional manner. Also, the interior of each column in the figure can be coded in different ways to form a stacked graph.

Advantages: Since naked eye is very sensitive to height differences, column charts can use the height of the columns to clearly reflect the differences in the data.
Disadvantages: Column charts are suitable for small- and medium-sized datasets.

Traditional column charts are generally used to display changes over a certain period or comparisons between different items, to indicate the comparison or change rule of the absolute quantity. Traditional 2D column charts include 2D clustered column charts, 2D stacked column charts, and 2D percent stacked column charts (Figures 5.4–5.6).

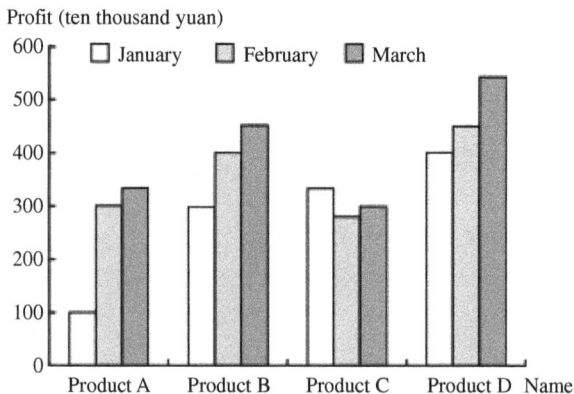

Figure 5.4   2D Clustered Column Chart

Sales revenue of product A and product B

Income (ten thousand yuan)

| □ Net sales profit A | ▨ Net sales profit B | ▨ Margin A |
| □ Margin B | ▨ Profit A | ▨ Profit B |

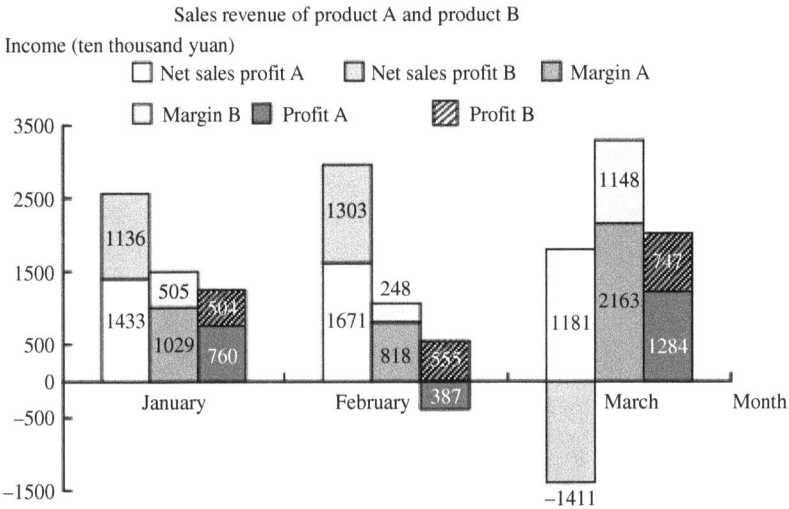

Figure 5.5   2D Stacked Column Chart

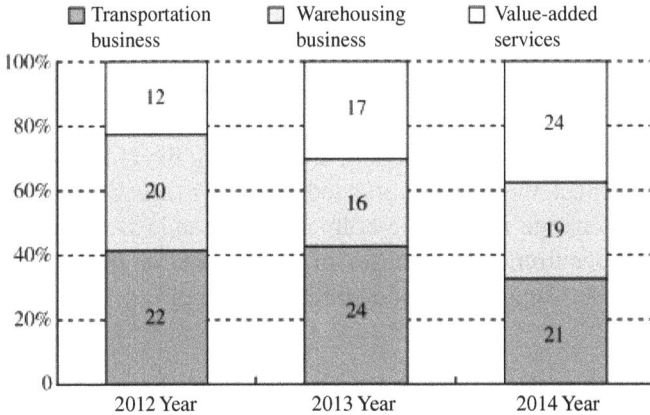

Figure 5.6   2D Percent Stacked Column Chart

In addition, column charts can also be shown in more intuitive three-dimensional charts form. The three-dimensional histogram can display three different variables on three axes, which is three-dimensional and visually striking. Here are three typical 3D charts: 3D Clustered Column, 3D Stacked Column, and 3D Percent Stacked Column (Figure 5.7).

Score

2010

(a)

(b)

(c)

Project 3
Project 2
Project 1

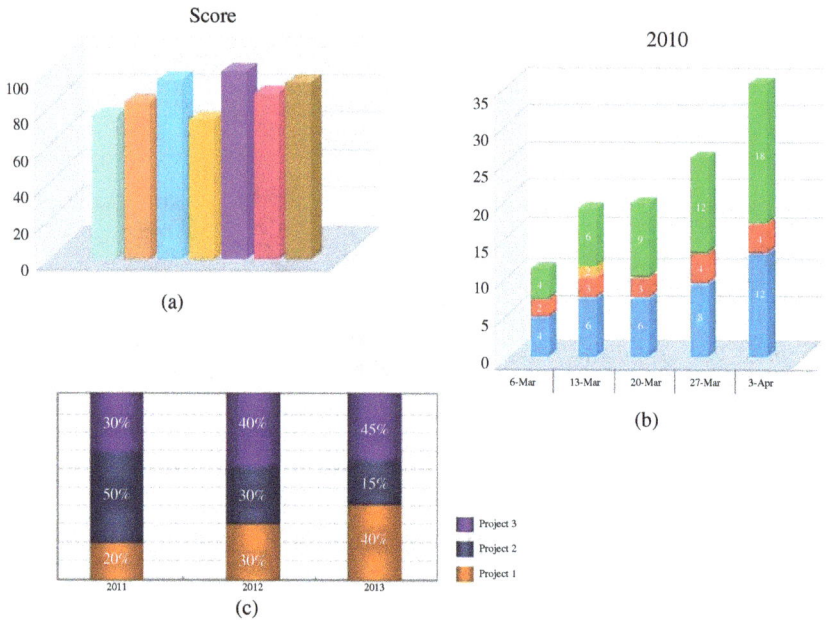

Figure 5.7   3D Column Chart, (a) 3D Clustered Column Chart; (b) 3D Stacked Column Chart; (c) 3D Percentage Stacked Column Chart

The general ideas for analyzing and viewing the column chart are as follows: (1) check the content reflected by the *x*-axis and *y*-axis; (2) check the size and change rule of the column value; and (3) comprehensively analyze the occurrence and causes of phenomena or problems, and put forward relevant suggestions and countermeasures.

Bar chart

The column or row of a worksheet can be plotted into a bar chart (Figure 5.8), which displays comparisons between items, allowing one to see the differences between the data immediately.

Depicting a bar chart mainly includes three elements: the number of groups, the width of the group, and the limit of the group. It is especially suitable for such scenarios where the axis labels are too long and the displayed values are persistent.

Average: 322.4000

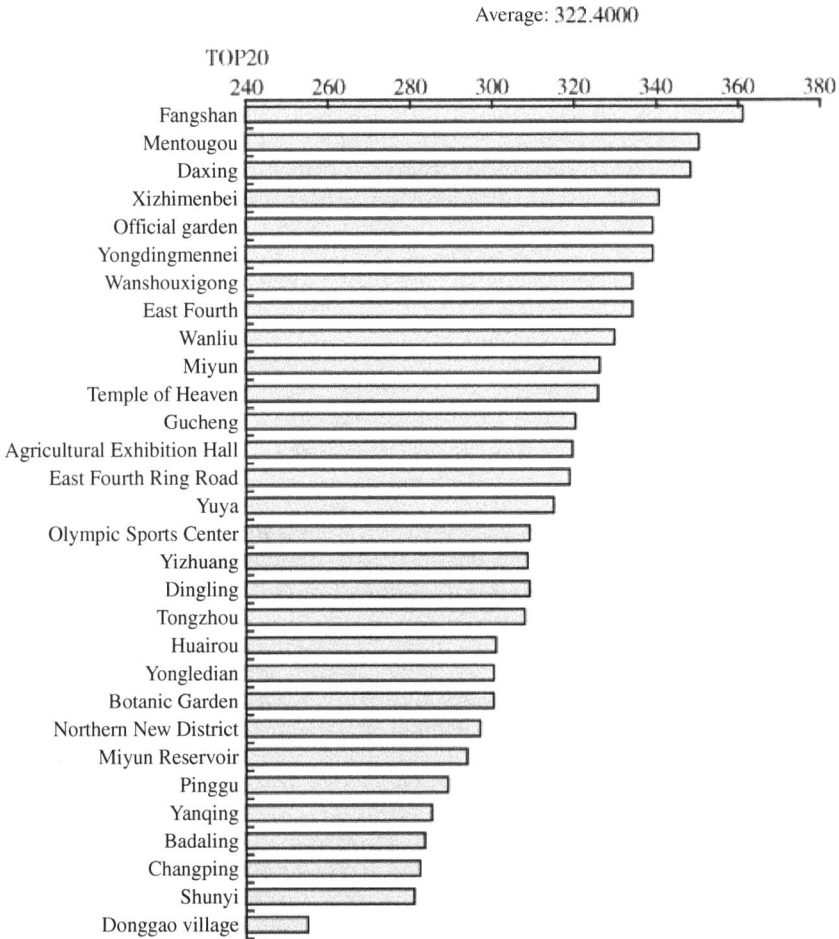

Figure 5.8   PM2.5 Data Detected by 30 Monitoring Stations in Beijing (December 20, 2016)

Source: China Industrial Information Network.

Figure 5.8 shows the PM2.5 detection data of 30 monitoring stations in Beijing. The *x*-axis and *y*-axis represent the value of PM2.5 and monitoring stations, respectively. From them, we can clearly and intuitively see the value detected by each monitoring station and its magnitude relationship.

Line chart

The line chart reflects the development trend and distribution of things with a combination of points and lines. It is suitable for expressing increases and growth values, two-dimensional large data sets, and especially for those where the trend is more important than the exact value. It is also suitable for comparisons between multiple two-dimensional datasets, and can display trends by time (year, month, week, and day) or by category when the order of many data points needs to be represented, as shown in Figure 5.9.

Pie chart

The pie chart is generally used for reflecting the proportion of a certain part, expressing the visualization of one-dimensional data. In particular, it can directly reflect the size, sum, and mutual proportion of each item in the data series. The individual data series in the chart are identified with different colors or patterns to represent how things are made up. Pie charts are most useful for comparing the percentage values of different data in the sum (Figure 5.10).

It uses different colors to distinguish the partial modules, and the share of the partial to the whole is clear at a glance, which can intuitively

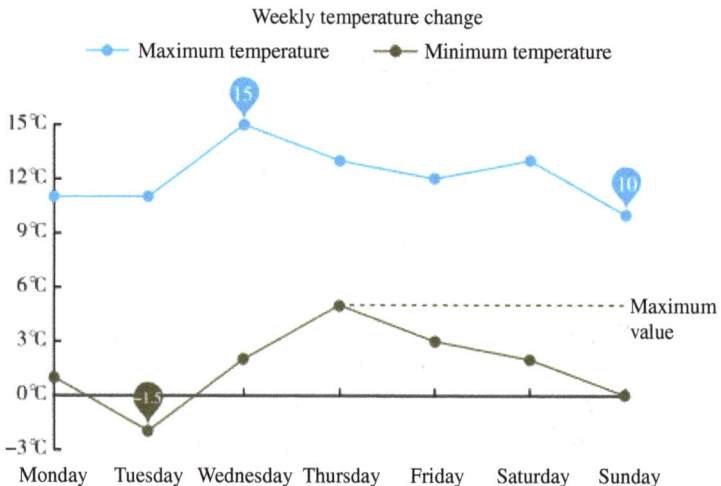

Figure 5.9   Line Chart

User access source of a site
- Direct access
- Email marketing
- Alliance advertising
- Video advertising
- Search engine

Direct access
12.5%

Email marketing
12.5%

Alliance advertising
10%

Video advertising
8%

Search engine
57%

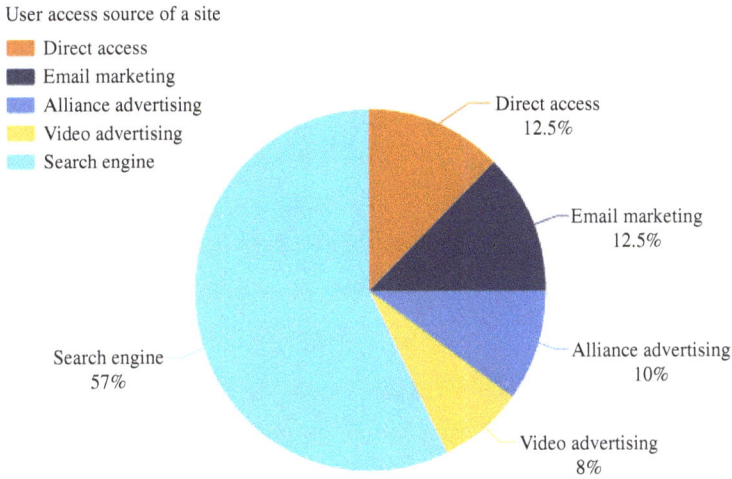

Figure 5.10   Pie Chart

and clearly reflect the proportion information. However, in pie charts, there is only one data series to be plotted, and values must be positive and mostly not zero, which brings limitations for fine data applications.

Scatter plot

Scatter plots show the relationship between pairs of numbers and the trends. For each pair, one number is plotted on the x-axis and the other number is plotted on the y-axis. The intersection of the x-axis and y-axis is marked. When a large number of data pairs are plotted, a graph appears (Figure 5.11).

Since the scatter plot can intuitively reflect the changing trend between variables, it is helpful for analysts to decide the analysis method to simulate and present, such as the function curve drawing. Meanwhile, it can also be used to draw average lines, assist in adding text labels, perform matrix correlation analysis, etc., which is often used in teaching and scientific computing.

Radar chart

Radar charts, also known as Debra charts and spider web charts, are mainly used to express the distribution of things in various dimensions. In

BMI of 507 sample individuals

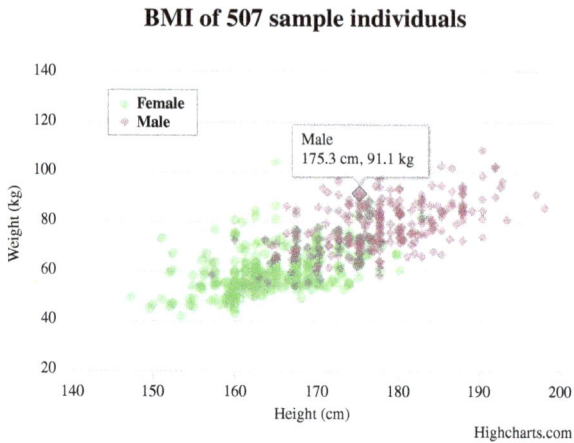

Figure 5.11   Scatter Plot

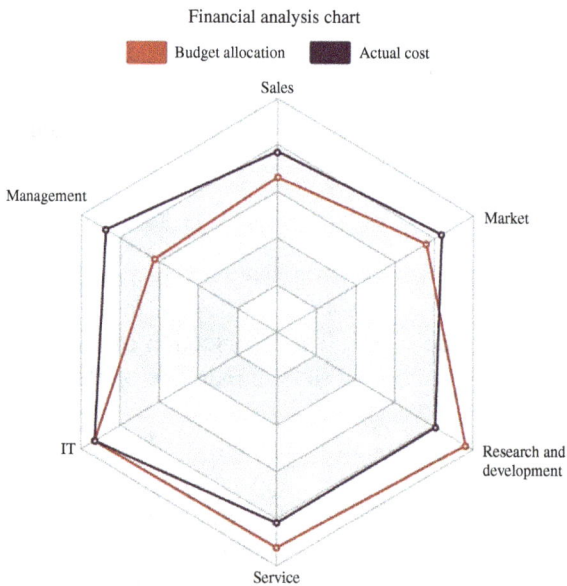

Figure 5.12   Radar Chart

terms of specific expression, the radar chart reflects the proportion of each individual on a circular chart, forming a diagram like a spider web, and providing an intuitive and eye-catching chart reference for data analysis and comparison (Figure 5.12).

Radar chart is suitable for multi-dimensional data (more than four dimensions), where each dimension must be sortable. It is often used in financial analysis reports, which can accurately represent the company's financial ratios to help understand the changes in the company's financial indicators and their good and bad trends.

Typical visualization tools

The statistical chart visualization methods were discussed above. When choosing the correct charts and graphs for data, one should also refer to the scientific visualization model (Figure 5.13), compare the advantages, and use multiple methods to jointly present the data.

When making multiple graphs, we should compare all the variables to see if there are any questions that are worth further study. Start by looking at the data in general and then zoom in to specific categories and individual features.

Selection method of basic charts: Compared with the traditional way of displaying data in tables or documents, data visualization is used to display data, making the data more intuitive and convincing. In various reports and descriptive documents, the data are presented with intuitive charts, which is more concise and reliable. Echarts can be used for the drawing of statistical charts for front-end development (Figure 5.14).

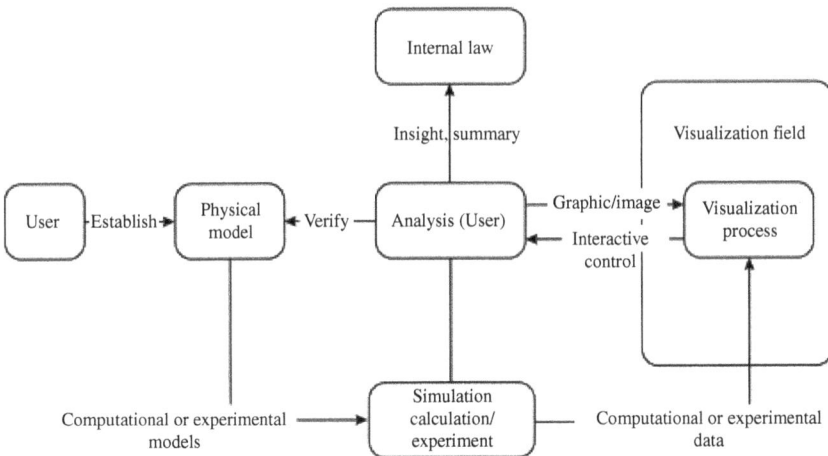

Figure 5.13   Scientific Visualization Model

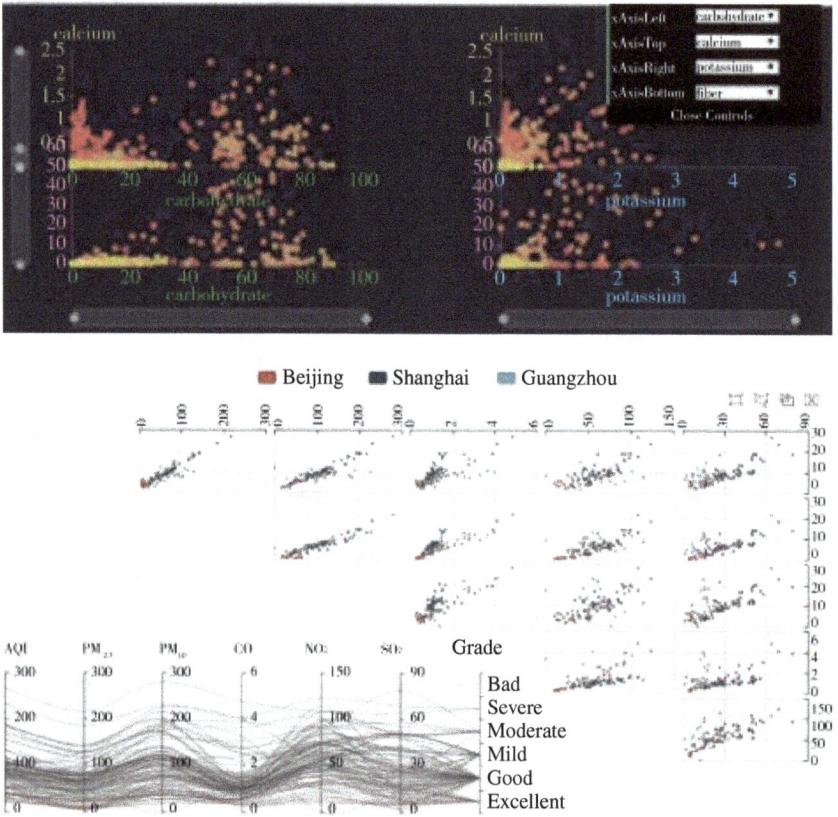

Figure 5.14   Echarts Visualization

Echarts is a pure JavaScript icon library that can run smoothly on PC and mobile devices, compatible with most current browsers (such as IE8/9/10/11, Chrome, Firefox, and Safari). The bottom layer relies on the lightweight Canvas class library ZRender, which provides intuitive, vivid, interactive, and highly customizable data visualization charts.

Currently, Echarts has released version 3, with more rich interactive functions and visual effects. Version 3 has also been deeply optimized for the mobile terminal. The complete documentation can be found on Echarts official website.

Chapter 6

Big Data Empowers Smart Government Affairs

On a global scale, big data empowering smart government has become a new form of future government development. China's *13th Five-Year Plan* clearly proposes "implementing the national big data strategy" to use big data to help the transformation of government functions and innovation in social governance. Smart government uses big data technology to integrate various related data information. It takes data analysis as the core, and makes intelligent analysis judgments and scientific decisions through multidimensional analysis and mining applications of massive data. This will help accurately catch the dynamic changes in government affairs, discover new needs of the public, change passive services to active services, and thus greatly empower government services. Based on this, an era of big data belonging to smart government affairs is on the way. This will change the existing government service model, promote the development of government service concepts and work processes, and lead the successful transformation and upgrading of government services.

Smart Government Starts with Big Data

The government is the most important decision-making body of the country. Whether its decision-making system is scientific or not, it directly determines the governance capacity and effect. China's *e*-government system involves various aspects, such as national livelihood, social security, traffic management, finance, environmental protection, medical care, and education. However, the existing system still has the problem of "information island" where most systems cannot share data with the outside world. Currently, they can only realize the informatization of the office process but not scientific decision-making based on data through data analysis. There is still a big gap between the current situation and the real "smart government".

What is smart government? The key of smart government affairs is "government affairs", with big data as its manifestation and carrier. Through big data technology, standard data transmission, storage, and processing can be provided in the heterogeneous environment of data sources and networks. Its inclusiveness can weaken the boundaries between government departments and between the government and the public, making it possible to share data, thereby improving the efficiency of government's works and collaborations and the efficiency of working for people. This could also improve the social capabilities of governance and public service. From a technical point of view, *e*-government and smart government are not an A- or -B relationship. They are different but closely related. Without the support of big data technology, the *e*-government system can only be regarded as the initial stage of smart government. Without the support of *e*-government system data and informatization, smart government will be like water without sources and trees without roots. The construction of the existing government affairs informatization has accumulated a large amount of data, and big data technology is essential to ensure the comprehensiveness and accuracy of government affairs data. It can make full use of the value of government affairs related to big data and deeply integrate the data with businesses so that the construction of smart government affairs will have substantial significance instead of just resulting in one more vanity project.

The key work of smart government is to make government data "speak" and "smart". For example, in the public security system, big data can form a comprehensive think tank for public security including criminal investigation, technical investigation, network investigation, and economic investigation, and provide visual analysis and business connection for public security police case handling; in food and drug supervision, big data can form a three-dimensional management of the whole process and life cycle of production, processing, inspection, logistics, and sales, thus effectively reducing transaction costs, promoting specialized division of labor, and improving work efficiency.

Therefore, smart government affairs based on big data can realize the intensification of government affairs services. There are many government service departments that are oriented to different service objects, subjects, and contents. To achieve the intensification of government affairs services, the first thing is to realize the cross-connection of multiple business systems through data sharing and integration, and simplify the government service process; the second is to establish standards and mechanisms for data sharing and integration to significantly improve the coverage and comprehensiveness of government data, thereby reacting on government services and effectively improving the service qualities. Big data transforms government services from passive services to active services, which greatly improves the customer experience of the main service objects such as the public and enterprises, and brings more convenience to the public.

How to Carry Out the Construction of Smart Government Affairs

In order to further complete the construction of the government affairs system and realize smart government affairs, we should start from three aspects: (1) complete the construction of *e*-government system, (2) promote data sharing among multiple departments, and (3) build a government system with smart functions. Based on this, we could achieve "speak with data, make decisions with data, manage with data, and innovate with data".

The construction of e-government system

In order to further improve the *e*-government system, the existing *e*-government resources should be integrated to share data among different departments, and should also enable the construction of the latter. At the same time, more data-sharing requirements should be applied to platforms that are planned to be constructed or have not yet been constructed. Among them, it is better to build some national big data platforms, such as demographic information, credit assessment, and other platforms closely related to the macro economy, where government departments at all levels upload relevant data to the platform. This can effectively avoid the waste of resources and can also establish the authority of the platform.

The construction of *e*-government system lies in the collection, classification, and preprocessing of data. For data collection, technical standards and rules and regulations should be established; the collected data should be classified accordingly so that it is clear whether data are classified, can be shared, or can be open to the public. Data preprocessing includes data cleaning, screening, labeling, and error correction. It should strictly follow the corresponding specifications, provide high-quality data that meets the standards, and provide strong support for subsequent data processing. Here, we suggest working with high-tech enterprises to jointly develop big data platform technology, promote the industrialization of big data, and strive to command heights of the development and application of government big data. Also, we should utilize the existing data advantages of government affairs to activate innovations in transforming government functions, strengthening social supervision, and reducing administrative supervision costs.

Promote data sharing among multiple departments

Data sharing among multiple departments is the basis for transforming government functions, strengthening social supervision, and promoting innovation and entrepreneurship. Since the government and public departments have mastered more than 80% of the social data resources, to realize the grid management and services, it is necessary to make full use of the existing data resources and realize data sharing among multiple departments.

The government data authorized by government system could empower the public services. Through these open data, the public can better understand the development of social economy and government informatization. It can also help the public through the construction of the *e*-government website platform to simplify the handling procedures and improve the efficiency of government services for the public. At the same time, through the administrative licensing and punishment data developed by the government affairs system, the public can better supervise government work, improving the transparency of administrative management and the credibility of the government.

Big data analysis can predict market demand, population flow, social and economic situation, etc. To achieve the above functions, there must be massive data resources. The government and the public departments hold a large amount of data resources, and it is possible to achieve sharing only by recommending data sharing among multiple departments. Through sharing these relevant data resources, it can save the time for social entrepreneurs to collect data, save the cost of starting a business, promote the development of social innovation and entrepreneurship, and provide a new model for social economic growth.

Building a government system with smart functions

Based on the sharing of existing data and the updated collection of new data, the use of big data technology to build a smart government system can not only achieve smart service functions for the public but also help smart management and decision-making. To form an *e*-government system with intelligent functions, we believe that it should at least be able to assist the government to improve the efficiency of supervision and service, improve the government information services, establish early warning of social and economic risks, and formulate relevant policies and regulations in a scientific way.

First, assist the government to improve supervision and service efficiency. The platform can analyze the operation of enterprises in different regions and industries, and accurately notice the current market weather vane, thus better guiding the business activities, providing services for enterprises more efficiently, and better realizing market supervision.

Second, improve the government information services. The platform can analyze and process social and economic data, and use the results as the basis for government departments to release information. Further, since it is social and economic data, the results can not only meet the needs of enterprises but government departments can also review the relevant credit data and publish it on a unified platform to provide the public with authoritative credit inquiry services.

Third, establish social and economic risk early warning system. The platform can integrate massive data from various departments and industries, and obtain the development and supply and demand status of various industries. Such macro data can be used to assess the risks of economic operation and social public opinion, and realize the statistics of sensitive words, thereby helping the government prepare for any event in advance, reduce risk, and maintain social stability.

Finally, formulate relevant policies and regulations scientifically. The platform can analyze macro data through big data and establish an analysis model for reference, which is conducive to the government to scientifically formulate relevant policies and regulations, and thereby provide better public services.

Smart Government Application Direction

Based on big data, cloud computing, and other information technologies, smart government optimizes the concept and method of governance, which reflects the trend of the government's public service model changing from omnipotent to intelligent. It is a new direction of *e*-government applications in the era of big data, that is, to provide the public with intelligent services through intelligent decision-making.

Data-driven smart office

In smart government, big data, cloud computing, artificial intelligence, and other technical means are used to upgrade the traditional *e*-government system to an intelligent system. Then, it is to gradually realize a three-dimensional, multi-level, and all-round *e*-government public service system. Finally, it is to accelerate the initial application of the new model of intelligent *e*-government services.

Improve office efficiency

In the era of data-driven intelligence, most of the intelligent *e*-government systems are capable of office behavior analysis. It can automatically optimize the user's interface and common functions according to the user's position, authority, frequency of use, completed work tasks, etc., and help users use the government system more conveniently through further optimization and adjustment. At the same time, it will also have automatic office reminders, including email reminders, meeting notification reminders, and important office event reminders. The user can know which events need to be dealt with through the automatic reminder of the system, and sort the to-do events, such as sorting according to the importance and urgency. Of course, in addition to the functions described above, it is convenient to inquire about policies, regulations, and work procedures, conduct mobile office anytime and anywhere, and share the work experience of others, which could also greatly improve work efficiency.

Refinement of office management

Big data can promote the government system to set up a detailed performance indicator system, conduct in-depth analysis of key and difficult indicators, make real-time and quantifiable performance measurement possible, and identify and deal with poor performance in a timely manner. It is conducive to optimizing the allocation of work resources, improving the office work's overall performance, and leading to more refined management of the government affairs system.

Coordination of government affairs

Data are an important support for conducting business. Big data technology can help the government to build a big data platform that spans multiple departments and systems, promote the cross connectivity, and realize data sharing and business collaboration of various departments. This could eliminate the information islands and break down information barriers, thereby greatly improving the efficiency of government office and work, and bringing convenience to the public. At the same time, this could reduce the operating cost and expenses of the government system.

High-level public participation

Smart government establishes a communication channel between the public and the government, allowing the public to participate in policy formulation and implementation, effect evaluation and supervision through online interaction, making government work more transparent. For example, the government affairs system can develop social functions such as Weibo and WeChat public accounts, and make use of its openness and interactivity to enable the public to participate in government affairs. The massive information fed back by the public is then processed centrally, and the analysis results are fed back to the government to solve the relevant problems. This will increase the public's confidence in participating in politics.

Personalized intelligent service

The application of big data makes personalized intelligent services possible, which stems from the nature of fine information granularity in big data. For example, by establishing a personal information database, the government can provide faster and more accurate information feedback to individuals. Under the premise of complying with national laws and regulations, the smart government system can access personal information data, open industry data, and social organization data into the data model for analysis. The analysis results can provide more diversified and personalized information services for all aspects of public life. For another example, the government office website can provide users with scenario-based services and guide users to handle related businesses.

Data-based scientific decision-making

In the era of big data, the government's decision-making no longer relies on experience and intuition, or flash of inspiration, but is based on scientific data analysis. The government should apply the big data way of thinking in government management and decision-making, and establish an intelligent decision-making system through data warehouse, data analysis, data mining, and other big data technologies. The system can automatically generate statistical reports and visual analysis according to business needs, visualize social development, economic development, government operation, etc., provide scientific basis for government decision-making, and assist leading cadres in making government work decisions. The system can also optimize government decision-making, track the effect of decision-making, and make government decision-making more accurate and reasonable.

Big data can not only provide scientific decision-making for the government but also provide corresponding decision-making feedback according to the needs of government affairs and the public. For example, Harvard University in the United States once opened free public courses to the world. Through public courses, Howard collected the information of students in the courses, analyzed the learning information of relevant countries, and then studied the learning behavior patterns of scholars from all over the world. Big data's decision-making mechanism, intelligent assistance mechanism, and feedback mechanism make government decision-making more accurate, enabling the government to provide the public with more intelligent services.

Intelligent supervision for long-term anti-corruption

In line with the anti-corruption principle of "prevention is better than cure", the most effective way for the government to manage corruption is to try to prevent corruption from the beginning. Smart government has unique advantages in preventing corruption, and can expand the role of anti-corruption to the government intranet, extranet, and the Internet. Compared with traditional anti-corruption methods, it has a wider impact and greater strength for anti-corruption.

Promote the transparency of the government affairs environment

Smart government affairs bring openness, interaction, equality, and transparency of the government affairs environment. It also accelerates the transformation of the government affairs environment from closed and mysterious to open and transparent, effectively restricts the office behavior of various government departments, and makes the middlemen who profit from corrupt behavior lose the market. Further, it could promote the transformation of the new government concept in the government system, and lead the spread of a diligent and clean ethos.

Promote the electronization of government functions to reduce corruption rates

Smart government affairs can promote the intelligent electronization and the institutionalized transformation of the technical level of government affairs, as well as forming the technical regulations. Such technical regulations are conducive to government departments to capture corrupt behaviors and reduce the breeding ground for corruption. At the same time, discretion can also be regulated to prevent abuse of power. Due to the development needs, the operation of government affairs has become regulated and responsibilities have been clarified. Therefore, the electronization of government functions also reduces the possibility of black box operations by internal staff, to a certain extent.

Promote barrier-free information transmission to reduce corruption

The rapid development of smart government affairs has revolutionized the information transmission methods of various government departments. The channels for information transmission between government and the public are more diversified, the amount of data exchanged is larger, and the information flow speed is faster. The scope of the audience has increased significantly, dredging the entire channel for information flow. The unimpeded transmission of information effectively reduces the intermediate links of information transmission. This changes the power operation from the dark box operation to the sunshine operation, enables the office staff and public to supervise the public power and the office, respectively, and thus minimizes the possibility of corruption.

Big Data and Government Governance

Big data plays a pivotal role in the development of smart government affairs and also in smart office, scientific decision-making, and long-term anti-corruption. It not only provides important data resources for government governance but is also becoming a powerful tool for state governance.

Case 1: Shenzhen Bao'an District Smart Government Service Platform

The construction of Shenzhen Bao'an District Smart Government Service Platform covers the construction of Shenzhen Bao'an District Government Service Center, 6 street government service centers, and 126 community government service centers, as well as the landing of Bao'an "administrative service hall, online service hall, all-round acceptance center, all-round consultation center", etc. It will be more convenient for Bao'an citizens to seek government services. Following the instructions of "a hall, a window, a website, a number", citizens can complete almost all matters.

"One hall" is the government service hall at all levels. All administrative approval services in the district will be included in the government service halls. Through the service model of front-end acceptance and back-end approval, it provides one-stop service for citizens and helps save processing time. "One window" means that all the windows in the service hall accept all kinds of business. Citizens only need to prepare materials and submit them to any window. There is no need to submit documents repeatedly or at many windows. "One website" means that the online service hall builds a convenient service network to achieve 100% online declaration and 100% online approval. "One number" refers to the 24-hour hotline service number officially launched by Bao'an, which is responsible for accepting consultations, complaints, opinions, and suggestions for citizens and enterprises within the functions of various departments of the district government.

The main highlights of this case are as follows:

100% Online Filing

Comprehensive Business Acceptance: Build a unified comprehensive acceptance and feedback system in Bao'an to uniformly receive applications from online, mobile terminal, and physical halls. The accepted items are automatically distributed to the background approval system for approval (Figure 6.1).

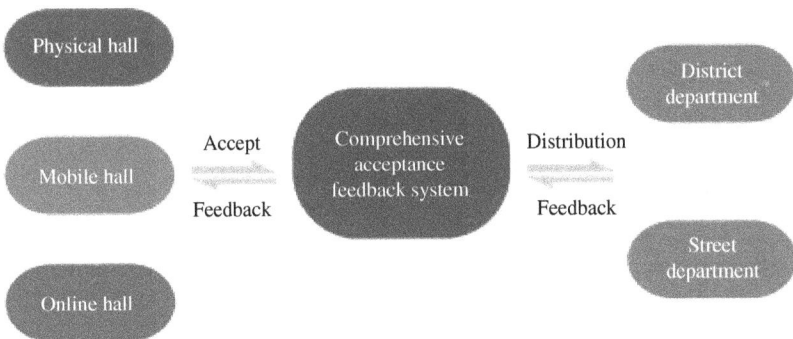

Figure 6.1   Business Process of Comprehensive Acceptance Feedback System

Integrate O2O Concept: Through adopting the method of "integration of online and offline", combined with "unified identity authentication, login, user space, license management", and other means, it can realize "one window for all works including receipt, comprehensive acceptance and feedback, certificate issuance, and all cases in the distinct could be processed nearby". At the same time, on the basis of "combination of physical hall and online hall", the mobile terminal application is expanded, and online declaration and material uploading are realized by means of mobile phones, tablets, notebooks, etc. (Figure 6.2).

Standardization of Administrative Approval: Through the standardized management system for administrative examination and approval, various links such as acceptance, examination and approval, and settlement are standardized, government service standards such as matters, data, processes, materials, and time limits are unified, and a unified handling process for the whole district is established. Relying on the "1+6+126" three-level government service center system and the comprehensive coverage of the smart government service platform, the integration of government services will be realized, and the "nearest handling" and undifferentiated "all-district handling" will be realized.

100% Online Approval

Full-Process Online Approval: The internal approval system of each department is connected with the whole-process electronic approval

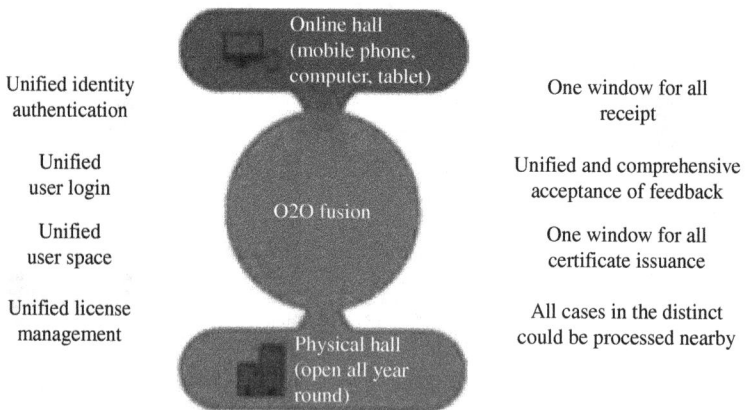

Figure 6.2   Online Hall Business Process

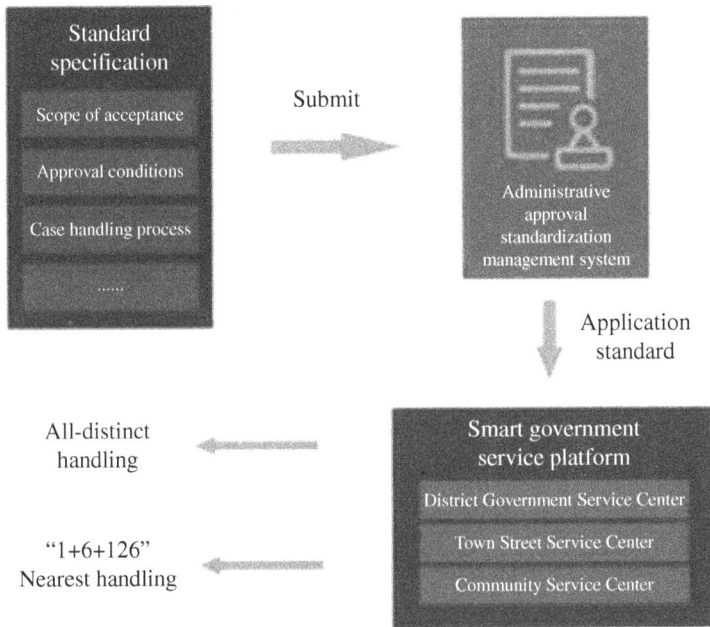

Figure 6.3   Standardized Management System for Administrative Approval

system according to unified standards. Using electronic signatures, material flow, license comparison, and other means to form a seamless collaborative mechanism of "front-end declaration, back-end approval, and license sharing". All internal audits, signatures, seals, license issuance, and other operations can be carried out online to achieve full-process electronic approval (Figures 6.3 and 6.4).

Material Circulation: We can use mobile phone photography, scanner, and other technical means to electronically digitize paper materials, and use electronic signature technology to verify the authenticity, integrity, and legality of submitted materials through the acceptance feedback system. Electronic official seal and other means are used for online approval, and the entire process of online declaration, online pre-approval, and other links is realized electronically.

Archiving: Relying on the approval archiving system, once approved, the application materials, approval materials, approval results, process record information, and other materials are archived and stored electronically, which lays the foundation for later data sharing and exchange.

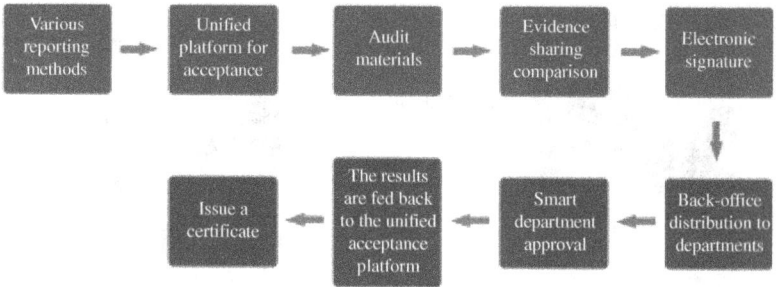

Various reporting channels

☐ Mobile phone ☐ Window ☐ Computer ☐ Self-service terminal

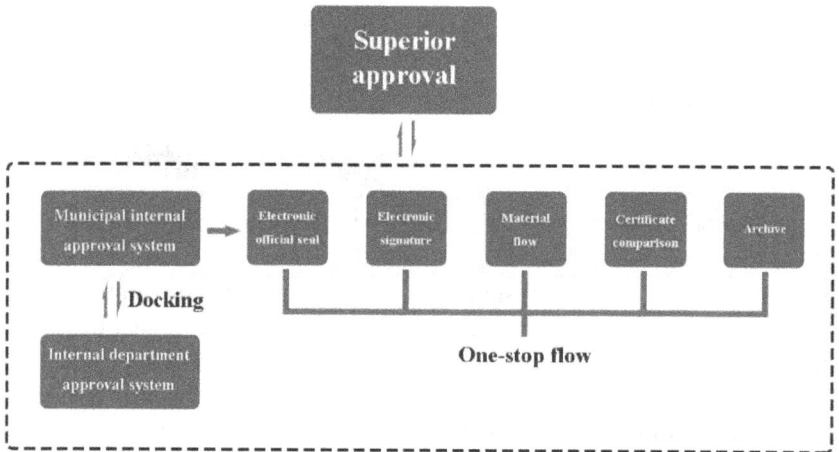

Figure 6.4   Full-Process Electronic Approval Business Process

Big Data Innovation Management

Real-Time Management and Control: Real-time management of "people, things, and windows" is realized through the virtual mapping of the physical hall, and the status of window service personnel. Through the smart government service platform, it is possible in real time to monitor the status of window service personnel, check the progress of approval items, and window distribution and dynamics, and realize the reverse tracking of the matters handled by specific service personnel, consequently building a supervision and assessment system.

Thematic Management: Through the accumulated business processing data, we can build personal space, enterprise space, and various thematic databases in Bao'an District. Based on the database, we can realize the traceability of users' online behavior data and automatically identify the historical data, thus achieving "only one-time fill-in is required for many times reuse and permanently share".

Integrity and Service: Relying on the Bao'an District Credit Information Platform, we can integrate credit data and build a large credit database. We can also include the untrustworthy behavior of individuals and enterprises into the integrity record, implement "trust approval", and create a green channel for "deficiency-tolerant approval". At the same time, through the implementation of "trust approval" and "deficiency tolerance approval", a whole-process management mechanism for credit information recording, collection, sharing, and application can be established, so as to promote various social entities to regulate and restrain their own behavior, and create a good atmosphere of integrity and law-abiding.

Case 2: Yan'an 12345 Smart Government Service Platform

In order to further expand public appeal channels, effectively solve practical problems, promote the efficiency of government services and improve the level of public services, and allow citizens to share the development results of smart government services, Yan'an Municipal Government has integrated various service hotlines for people's livelihood in the city and established the Yan'an 12345 Smart Government Service Platform.

Yan'an 12345 Smart Government Service Platform is an online and offline service integration window developed by the Yan'an Municipal Committee and Municipal Government for "the No. 1 leader accepts the matter, and the task is transferred according to the accusation". It integrates the public affairs and inquiries into one, acting as a "bridge" between the citizens and the government. On the one hand, citizens can seek help, consult information, supervise government work, and express opinions and suggestions through the 12345 platform; on the other hand, through public opinion surveys on social conditions, the new model of big data analysis provides reference resources for decision-making and work development. Yan'an 12345 Smart Government Service Platform was launched on July 28, 2017. It provides 24-hour manual service throughout the year. At present, the platform has received a total of 58,450 calls from citizens, and accepted 13,872 appeals, with the settlement rate and mass satisfaction rate of 97.31% and 86.64%, respectively.

Platform Features

(1) *Orderly integrate the hotlines of various departments*: The integration of service hotlines such as the citizen hotline 2169000, the municipal office 12310, the provident fund 12329, and the tourist complaint 8013939 has been completed, which can realize the voice conversion and data exchange between the public security and emergency hotline platforms.

(2) *Expand and extend the acceptance channels*: In addition to the hotline, the 12345 smart government service platform has also expanded five acceptance channels, including Weibo, WeChat, email, SMS, and fax, forming a "six-in-one" multi-channel acceptance method.

(3) *Strengthen policy innovation propaganda*: (a) Improve the work system, and clarify the acceptance principles, handling responsibilities, handling time limit, punishment measures, and other regulations. (b) Sort out the responsibilities and boundaries of the units, and divide 20 categories and 60 sub-categories into a total of more than 800 items of division of labor. (c) Establish a "monthly notification and monthly supervision" system to ensure that citizens' demands are resolved in a timely manner. (d) Set up a hotline for return visits, met the satisfaction of the citizens with the events they appeal to, and supervise the organizers. (e) Innovatively start the "12345 Echo Wall" to publicize the results of typical cases. (f) Publicize through TV programs such as "Civilization Yan'an" and "People's Livelihood Yan'an", and cooperate with radio stations to broadcast appeal cases, on-site acceptance of citizens' needs, and on-site Q&A. The overall awareness of the 12345 smart government service platform was further improved.

(4) *Enrich information sources for government decision-making*: Yan'an 12345 Smart Government Affairs Service Platform can carry out social opinion surveys through standardized processes such as on-site observation, on-site recording, on-site telephone interviews, and on-site score calculation, and then feed back the most direct and true survey results to relevant departments, giving full play to its auxiliary decision-making role. As of May 2017, 6 social opinion survey projects have been

completed, 335.7 million valid samples have been obtained, and 4 survey reports have been formed.

Technology Implementation
The technical architecture of Yan'an 12345 smart government service platform is shown in Figure 6.5.

As shown in the Figure 6.5, the data aggregation layer realizes the unified aggregation of all data and builds the data resource pool of the big data platform. The data calculation layer analyzes and processes the data through index calculation, text analysis, and other methods to mine valuable information. The data application layer provides predictive analysis, report analysis, decision analysis, and other functions through the

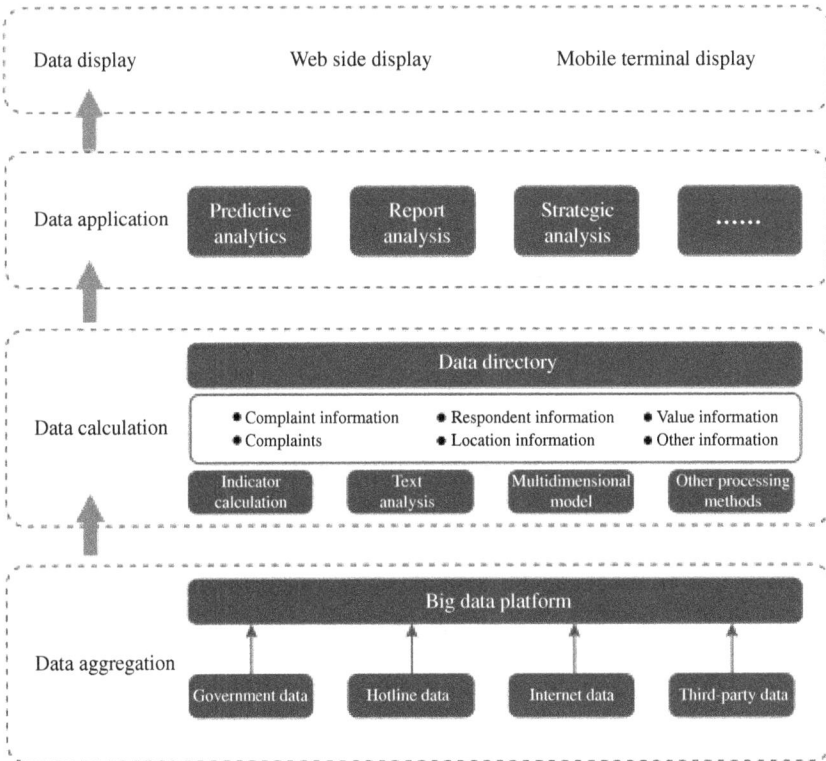

Figure 6.5   Technical Architecture of Yan'an 12345 Smart Government Service Platform

platform. Lastly, the data display layer mainly displays the application results, which can be displayed on the web and mobile terminals.

In the technical architecture, the big data analysis and processing in the data computing layer are the core. In the 12345 hotline, the content of citizens' appeals is basically text data. The data are first divided into words and get labeled according to data attributes or keywords, followed by automatic classification and rule classification. Finally, the data analysis and processing results are displayed to end users.

Application Effects

(1) *Real-time early warning of hot events*: The big data analysis of the 12345 smart government service platform is more accurate and can quickly reflect the trend changes of citizens' demands. Through the statistics, analysis, and processing of citizens' demands, we can sort out the sensitive information concerned by government departments, and provide real-time early warning to hot issues and emergencies. At the same time, combined with external data, it can conduct public opinion analysis, predict the event trend that may be found, and push it to relevant departments through websites, emails, text messages, etc.

(2) *Data improve service quality*: By analyzing the frequency of the keywords mentioned by the public at 12345 hotline, it is possible to determine the hotspots that public pay attention to in a certain period of time, so as to know what the public care about. Thereby, the government could solve problems in a targeted manner, such as public response through websites and information push.

(3) *Promote the improvement of government management*: The 12345 hotline uses big data to analyze hot issues, especially for service quality, work efficiency, work style, and hotspots and difficulties in the industry that the public report on a regular basis, making the 12345 hotline an important means to improve the overall management level of the industry. For example, to monitor the handling of each undertaking unit, it uses color signs to indicate changes in the time limit, and generates warning lines for early warning and prediction. For another example, considering indicators such as the number of supervisions, extension time, public satisfaction, the number of refunds, and the number of complaints, it can reflect the problems existing in a certain area or department in a certain period.

Future Development Path

There is no doubt about the importance and necessity of developing smart government affairs. In the specific planning and development, we need to strengthen the top-level planning and design, implement the construction purpose of "serving the people", and promote the transformation of governance thinking.

Strengthen the top-level planning and design of smart government affairs

The construction of smart government affairs has certain complexity and continuity that cannot be achieved in one step. Detailed construction and implementation steps are required. The government needs to gradually promote the construction of smart government affairs according to the local development status. In particular, the current smart government affair is in the stage of practice and exploration, and it is necessary to strengthen the top-level planning of smart government affairs and improve the system design. First, we must clarify the construction goals, and then establish the overall strategy, planning, and specific content of smart government affairs. Second, we must strengthen resource integration and information sharing, and promote the interconnection of data and business among departments. Finally, we need to promote the construction of government information disclosure mechanism, where any government information must be legally and safely disclosed.

Implement the construction tenet of "serving the people"

We must strengthen bottom-up public supervision and management while clarifying the top-level planning. In the meantime, we should also rely on big data analysis and processing technology to process and feedback government big data. These can integrate social and national resources for governance, expand the scope and channels for the public participation in accordance with the law, and help strengthen the public's supervision of government functional departments. With the circulation of government affairs information between the public and the government, the government can obtain information on public demands and supervision at any time. This can make the government work more efficient and transparent, make the construction content and construction goals practical, avoid the "vanity project", and thereby could serve the people and the society better.

Promoting a change in the thinking of governance

The core of smart government is to change the current social governance model, and the most important thing is to change the way of thinking of government officials. This is the development trend of future governance, and it is the basic condition for transforming governance thinking and building a service-oriented government. By integrating advanced technologies such as cloud computing, big data, and the Internet for government affairs applications, government affairs services could be more intelligent and flexible, and the social governance could be more detailed and scientific. This is of great significance for comprehensively promoting modern social governance and building a harmonious society.

Chapter 7

Big Data Boosts Economic Growth

Big data has become a national strategy that plays a pivotal role in the economy. For this reason, local governments keep pace with the times, adapt measures to local conditions, locate themselves on regional positioning, rely on information sharing and data analysis on online platforms, and gradually integrate big data applications with investment promotion, targeted poverty alleviation, industrial transformation, intelligent manufacturing, data drainage, service upgrades, and smart finance closely, thereby maximizing the value of big data.

Accurate and Targeted Investment Promotion

Investment promotion is one of the key strategies to speed up industrial restructuring and upgrading. In the past years, we used to attract investment through tax incentives, supporting infrastructure, and public services, which has indeed stimulated the rapid economic development in various places. Under the impact of a new generation of technologies such as big data and artificial intelligence, traditional preferential policies such as taxation, land, and "three access and one leveling" (access to water supply, electricity, and roads, as well as land leveling) are no longer the only investment promotion measures. How to apply computing power, storage capacity, and deep mining based on massive data to provide a reference for accurate investment promotion has become an important issue in investment promotion.

At present, big data has gradually become an important technology for the government's refined management. In terms of investment promotion, it is urgently needed for investment promotion departments such as local development and reform commissions, investment promotion bureaus, and industrial park management committees to apply big data to innovate investment promotion modes and means. Then, they should use these modes and means to expand investment and investment channels. Moreover, this helps deeply mine investment resources and improve the efficiency and quality of investment based on enterprise information, transcending time, space, and geographical constraints.

Gathering big data for investment promotion

At present, more and more government management departments have begun to obtain panoramic images and dynamic investment information of various enterprises by establishing an investment big data service platform that integrates core data capture, mining, and analysis. On the one hand, these data come from the real-time capture of Internet data websites, i.e., through the data crawler and real-time capture of the data of enterprises and ministries. On the other hand, it also covers the data obtained by the cooperative industrial park through data declaration and data collection. Based on the two means, a detailed investment database can be established.

In this database, the management department can quickly check the panoramic information of the enterprises in the database, including the basic information of enterprises, investment information, honorary qualifications, intellectual property rights, talent recruitment, annual report information, transaction information, reward, and punishment warnings, to establish a three-dimensional corporate portrait. At the same time, through the enterprise portrait, it could quickly find the relationship between the investment, senior management, patents, bidding, litigation, etc. of the target company, explore layer by layer of a company to form a complex relationship network, and intuitively and stereoscopically display various associations in their business activities. This could be used to accurately describe the development layout of enterprises or discover enterprise group risks.

Therefore, this investment database is like a "treasure map" for precise investment promotion. Managers can quickly follow the map to achieve "precision guidance". For example, the People's Government of Yaohai District, Hefei City has previously focused on companies with a registered capital of more than 5 million Chinese Yuan, and analyzed their industrial types, formats, investment sources, business scope, and other elements. They then analyzed the companies from the aspects of market development and industry prospects, combined with the industrial development direction of the district. Finally, the Yaohai District was able to identify the key entry point of the industrial development of the district and the investment needs of enterprises, as well as the companies closely related, and the precise introduction of the project was realized.

Explore the investment target

Simple collection and integration of data provide very limited value. Only by completing further analysis and mining of data and information, and extracting effective information from massive data, can big data play its due value. In the process of attracting investment, based on the investment database mentioned above, one can establish an appropriate economic model, evaluate the development status of enterprises, effectively get the matching degree of local resources and enterprises, and conduct demand analysis for specific investors. These can help the government and corporate investment promotion departments to accurately select potential investment companies.

Meanwhile, the platform can provide recommended labels for the government through industrial information classification and labeling, and through matching technologies and algorithms to realize accurate matching. As a result, with the support of big data, the time it takes for government departments to select suitable targets and tailor-made investment plans is greatly shortened, and the cost of finding investment leads is reduced. In this way, the efficiency of investment promotion can be quickly improved, the implementation of projects can be promoted, and the government can achieve a seamless connection with projects, capital, talents, and the market.

Implementation of investment promotion channels

In order to make the investment database fully function, and to facilitate the successful entry of the precisely selected target enterprises into the investment promotion process, it is necessary to develop the investment promotion path. The key to precise investment promotion lies in the uniqueness and certainty of the target enterprise. Therefore, it is necessary to keep an eye on the target enterprise, understand the basic information of the unit through the database, and clarify its fit with the investment promotion project through on-the-spot investigation.

On this basis, the investment promotion department needs to accurately plan major projects according to the actual situation of the specific target companies, as well as the previous data collection and processing. Then, it can make tailor-made investment promotion plans for the target companies, detailing the project overview, supporting conditions, policy support, and the needs of target companies, and send a small team to come to the company to attract investment, so as to better complete the two-way connection.

For the above trilogy of investment promotion, taking Gui'an New District in Guizhou Province as an example, a more intuitive understanding can be established. Gui'an New District has completed the construction of "six clouds", including approval cloud, regulatory cloud, supervision cloud, investment promotion cloud, license cloud, and analysis cloud, realizing the "gathering, generalization, and application" of data resources. Through cloud construction, high-end projects and real applications are connected, and the data are thus well used to serve government affairs management and local development. In particular, on the investment promotion cloud, the investment promotion department can quickly obtain various enterprise information.

Based on the investment promotion database, the person in charge of investment promotion could use data analysis technology and algorithms to accurately search for target companies. Also, through tailor-made investment promotion plans, the investment promotion can be quickly implemented. For Gui'an New District, a national-level new district, it has both demand and advantages in terms of new energy vehicles. Therefore, when Gui'an New District found out about the investment needs of Beijing Dianzhuang Technology Co. Ltd., they got in touch with the enterprise and formulated an investment promotion plan accordingly.

Finally, through multiple communications, Gui'an New District successfully signed a strategic cooperation framework agreement with Beijing Dianzhuang Technology Co. Ltd. The agreement stipulated that Beijing Dianzhuang would invest 10 billion Chinese Yuan in Gui'an New District from 2016 to 2018 to build the "Internet + New Energy Vehicle Full Ecological Industry Chain Cooperation Project". In this process, the specially designed investment promotion project played an important role, which directly promoted the landing of the whole plan.

This shows that the current investment promotion model is changing. Based on data collection and analysis, it could transform from the previous investment promotion model of "casting a wide net" to precise investment promotion. The investment promotion mechanism, investment direction, investment promotion measures, and investment promotion team are all in the transition to "precision". Consequently, the investment promotion department can also focus on the cutting edge and achieve the real purpose.

Industry Transformation and Intelligent Manufacturing

Industry is the key to developing a city. At present, the penetration of big data into traditional industries is accelerating. Under the influence of financial industry, manufacturing industry, service industry, and other industries, the production mode and management mode are constantly changing, and are gradually developing in the direction of networking, digitization, and intelligence. While realizing industrial transformation and upgrading, industrial elimination and emerging industries emerge one after another.

In the process of industrial transformation and upgrading, on the one hand, we can use big data to help enterprises transform from extensive development to refined development; on the other hand, it can further help industrial integration, rely on big data, and focus on the integration application of big data and cloud computing, Internet of Things, wearable devices, intelligent service robots, artificial intelligence, and virtual reality.

As far as the manufacturing industry is concerned, it can be described as the main battlefield for big data applications. According to the definition of the "Intelligent Manufacturing Development Plan (2016–2020)", intelligent manufacturing is based on the deep integration of a new generation of information and communication technology and advanced manufacturing technology. It is a new production method with functions, such as self-perception, self-learning, self-decision, self-execution, and self-adaptation throughout all aspects of manufacturing activities, such as design, production, management, and service.

In "Made in China 2025", the action plan for the first decade of China's implementation of the strategy of building a manufacturing power, intelligent manufacturing is listed as the main direction, where big data plays a pivotal role.

On one hand, big data personalizes the way of production. In the traditional production model, the enterprise is the dominant player, while the users buy the products and services that the enterprise provides. Now, with the support of big data technology, production models and business models tend to be customized. Combined with internal and external data, customer feedback obtained from social data is integrated into new product development, with the object and scale of enterprise production being determined accordingly, that is, consumer preferences and demands based on consumer data are forcing multiple links on the supply side of the manufacturing industry, including product design, R&D, production, supply chain, and marketing.

Among them, in terms of manufacturing industry data, it can be divided into two categories: internal data and external data. Internal data mainly includes operational data, operational data, customer data, product design data, R&D, and production data. External data includes social data, partner data, and Internet business data. In the application of manufacturing industry data, it is necessary to achieve the connection of internal and external data, realize the integration and standardization of internal data, and avoid the phenomenon of information islands.

On the other hand, big data can also achieve smart R&D and design, fast manufacturing, transparent production management, and after-sales service-oriented product. With the gradual construction of intelligent manufacturing, thousands of sensors will be installed on the production line in the later stage to monitor various production parameters. Adjustments through the application of big data technology will significantly improve production efficiency and product quality, and reduce production costs. Further, the integration of big data and supply chain will more clearly get the inventory content, order completion rate, material, and product distribution, and thus improve the response speed, reduce costs, and optimize inventory.

At present, in terms of intelligent manufacturing, many well-known enterprises in China and other countries have come to the forefront. For example, under the guidance of "Made in China 2025", Haier has gradually explored an intelligent manufacturing development route centered on interconnected factories, and established a modular, automated, digital, and intelligent manufacturing technology system. For instance, in response to modular requirements, a refrigerator originally had more than 300 parts, but now it is integrated into 23 modules on a unified modular platform, and the integration and innovation of generalization, standardization, and personalized modules can meet the individual needs of users.

Regarding automation, Haier focuses on realizing intelligent automation interconnected with users, that is, automated and flexible production is automatically driven by users' personalized orders. For digitalization, through the integration of five major systems with iMES as the core, the integration of the Internet of Things, the Internet, and the Internet of Services, as well as the people–people interconnection, people–machine interconnection, machine–things interconnection, and machine–machine interconnection, was realized. Ultimately, the product is made more intelligent, and the entire factory becomes a system as intelligent as the human brain, which automatically interacts with people, meets user needs, and automatically responds to users' personalized orders.

In foreign countries, the production of self-driving giant Tesla vehicles has already been automated and digitalized, which is one of the representatives of intelligent manufacturing. In Tesla's workshop, from raw materials to the assembly of finished products, almost all production processes and work (except for a small number of parts) are self-sufficient. That's why it is regarded as the most intelligent fully automated production workshop in the world. Also, in the four major manufacturing links, namely the stamping production line, the body center, the paint center, and the assembly center, there are at least 150 robots involved. Here, you can hardly see people.

Overall, data are critical to the entire process of a manufacturing company. From product design to product research and development, based on the analysis of consumer consumption behavior, product design and research and development needs can be positioned; based on the analysis of consumer preferences and demand for products, precise marketing can be achieved, product marketing plans can be determined, and product inventory can be effectively controlled.

At the same time, the realization of intelligent manufacturing is a gradual progress. The real intelligent manufacturing covers multiple links such as product design, R&D, production, logistics, marketing, and customer relations, among which cloud computing, big data, the Internet of Things, artificial intelligence, and other technologies are the key technologies to realize the whole process of data, and also the foundation of intelligent manufacturing.

Therefore, the application of big data to promote innovation in the manufacturing industry can start from the following aspects: (1) emphasize the importance of the construction and sharing of data resources; (2) value the importance of the development of data analysis algorithm models; (3) improve value-added service capabilities; (4) stick to the application orientation of people's livelihood; and (5) use accurate analysis of big data to understand consumer needs and service focus, and ultimately achieve mutual benefit and win-win between enterprises and consumers.

Data Drainage and Service Upgrade

In the commercial application, how to dig consumer demand and therefore efficiently integrate the supply chain to meet their needs has become a key factor to gain a competitive advantage. Data-based operations have gradually become the mainstream. Relying on the "genes" of big data, "people, products, and fields" can be redefined. Specifically, based on the analysis and processing of sufficient data samples (including purchase situation, consumption preference, user activity, platform data and traffic, and information feedback), we can investigate keywords and attributes, accurately depict user portraits, and thereby conduct targeted data drainage and further achieve service upgrades (Figure 7.2).

User portrait, i.e., the labeling of user information, is a virtual user abstracted by enterprises after collecting and analyzing user data, which can be considered as a virtual representative of real users. The core job of user portrait is to match the user with matching tags, and to construct and describe the user's characteristics through multiple dimensions, including the user's social attributes, living habits, and consumption behavior, and then reveal the user's personality characteristics.

With user portraits, enterprises can truly understand the needs of users, and thus it is possible for them to accurately locate users. On this basis, relying on modern information technology to establish a personalized customer communication service system, and push products or marketing information to specific user groups, could not only save marketing costs but also maximize marketing effects.

For example, the companies could crawl user data from WeChat, Weibo, Toutiao, and other apps, as well as various activity data, online

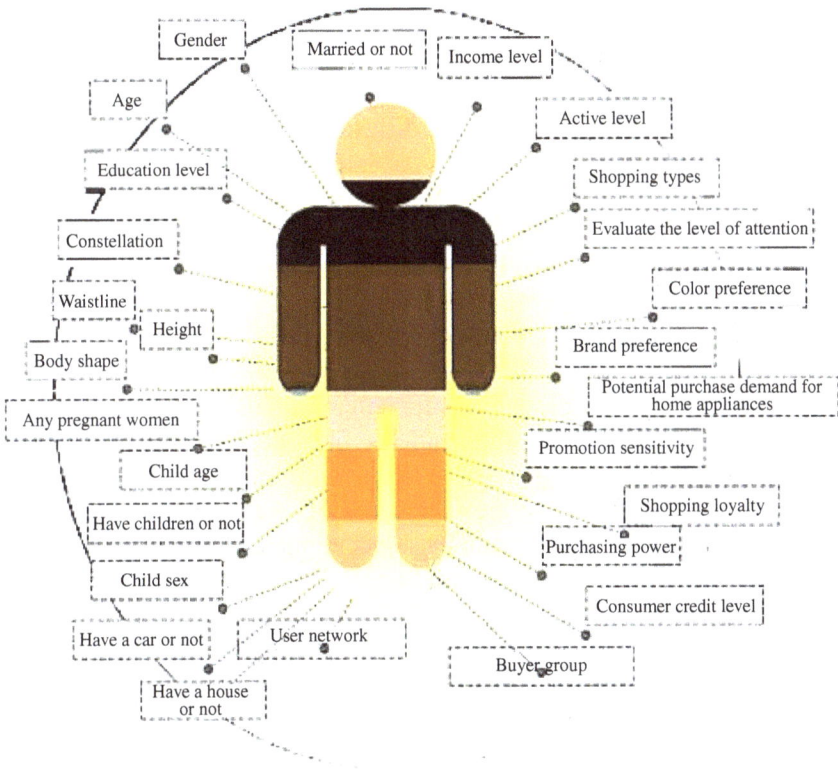

Figure 7.2   User Portrait

databases, customer service information, etc., and then clean and sort the data. Next, they could analyze the sorted data through classification, clustering, regression analysis, correlation analysis, etc., to further clarify users' interests, occupations, hobbies, and other labels. Based on this, they could realize user labeling and build user portraits through data modeling for data visualization analysis.

This is especially useful in the retail industry. On the one hand, through big data, it is possible to understand customers' consumption preferences and trends, send recommendation information for different products, carry out precise marketing of products, and reduce marketing costs. For example, it records the shopping habits of customers and reminds them to restock in time through precise advertisements before running out of daily necessities. On the other hand, according to customers' consumption behavior, we can recommend other products. For example, through customer purchase records, we can understand customers' purchasing preferences for related products, and put laundry-related products, such as laundry detergent, disinfectant, softener, and collar net, for sale together, thereby increasing the sales of related products.

Another typical case is to use store user portraits to boost data drainage and service upgrades. Through the labeling of store user information, the store can perfectly summarize the consumption characteristics of a user after collecting and analyzing the data of consumers' social attributes, living habits, consumption behavior, and other main information.

Store user portraits provide a sufficient information base for precision marketing, which can help stores quickly find accurate user groups, and analyze and dig store user needs. When the information is pushed based on the needs of store users, the acceptance of store users is maximized, which not only expands the amount of communication but also greatly increases the store's conversion trading volume, thereby improving the store's sales performance.

So how to capture store user portraits? We all know the importance of establishing store user file management. Through file records, we can collect and understand store users' consumption behavior, consumption preferences, consumption habits, consumption ability, and other related data information. This helps to quickly target targeted users, dig their needs, make more accurate marketing direction, and maximize benefits.

The key to the store user portrait is the label. What is a label? Take vehicle maintenance as an example, what age group your store users are in, what kind of car care methods they like, what brand of accessories they prefer, etc. If you know this information, you can label them accordingly.

Tags can be divided into the following categories: (1) Basic attributes, such as name, mobile phone, residence place, workplace, company, family life, circle of friends, personality, and other basic information. (2) Basic information, such as consumption proximity. Consumption recency refers to the time of the most recent consumption at the store, and divides store users into active users and sleeping users. (3) Behavioral preferences, such as door-to-door time, access method, category preference, and brand preference. The access mode and time period data analysis can be used to decide which way to push advertisements to store users.

(4) Consumption preferences, consumption periods, and preferred brands of store users. If some car owners like to wash their cars on weekends, you can set up corresponding packages in advance for customers, and guide the car owners immediately after entering the store, which can not only improve efficiency and joint rate but also make users feel intimately and privately served. (5) Store user services, such as store user membership level, evaluation level, complaint record, and return and exchange amount. (6) Business scenarios, labels such as store user DNA and store user privileges. When car owners and users with relatively large contributions enter the store, the store manager greets them to show respect. They can also provide some high-end services, such as free car washes and door-to-door services.

Finally, use labels to model store users, including three elements: time, location, and people. Simply put, it is what store users do at what time and in what place.

If you can make good use of the store user portraits, you can accurately locate the target users, catch the needs of the store users, dig the consumption potential of the store users, and carry out more effective precision marketing and personalized user services, thereby bringing more store entry rate and transaction rate to the store and improving the overall profitability of the store.

In addition, *e*-commerce can be said to be the first industry to use big data for precision marketing. The recommendation engine in the *e*-commerce website is based on the historical purchase behavior of customers and the purchase behavior of similar groups. The recommended product conversion rate is generally 6–8%.

Risk Control and Smart Finance

In the financial field, big data enables smart finance. Smart finance is based on cloud computing, big data, mobile Internet, artificial intelligence, and other technologies, which has comprehensively improved the business process, business development, and customer service in the financial industry, and realized a new financial business model of financing, investment, risk control, customer acquisition, and service. Smart finance is not only a simple combination of the Internet and the financial industry but also an emerging field derived from the combination of the traditional financial industry and the spirit of the Internet.

The three most important elements of smart finance are platform, data, and finance. At present, the market is not just about platform competition. In recent years, the Internet has shown explosive growth in finance, and has formed a pattern of integration of platforms, data, and finance. Only through big data can we connect platforms, data, finance, and other aspects, and whether the use of big data is reasonable is the key to the future market data battle.

The application direction of big data in smart finance

Financial enterprises are the forerunners of big data. Long before the rise of big data technology, the amount of data in the financial industry and the exploration of data applications have already involved the scope of big data. Nowadays, with the deepening of the application of big data technology, the concept of big data is gradually gaining popularity. While retaining the technical capabilities of the original data, financial institutions can effectively improve in many aspects such as customer management, product management, marketing management, system management, risk management, internal management, and optimization within financial enterprises through the effective integration of internal traditional data and external information sources. Next, we will introduce several typical application directions of big data.

Financial anti-fraud and analytics

Under the impact of the Internet economy, various terminals and channels are often facing various attacks. With the Internetization of banks, banks are facing severe challenges when developing various online financial innovation services. However, most current fraud analysis models are only able to detect fraud attempts and acts, yet the identification of potential fraud signals is often ambiguous.

In this regard, financial enterprises can build accurate and comprehensive anti-fraud information bases and anti-fraud user behavior profiles by collecting multi-directional data source information. Combined with big data analysis technology and machine learning algorithms, it could analyze and predict fraud behavior paths and effectively identify the fraud trigger mechanism. At the same time, through cooperating with business departments, it provides anti-fraud operational support and helps banks build a fraud information database. Eventually, this could help banks to predict the fraud in advance, accurately obtain the fraudulent path, and greatly reduce the losses.

Build a more comprehensive credit evaluation system

How to carry out risk control has always been the focus of the financial industry and one of the core competitiveness of financial enterprises. A fully functional credit evaluation system can not only effectively help financial enterprises reduce the cost of credit approval but also effectively control credit risks. When building a credit evaluation system, it should not evaluate whether a customer can borrow or how much money can be borrowed based only on loan standards. Instead, it must combine external transaction information and go deep into the industry to evaluate it with industry standards. Big data technology helps financial institutions establish a more efficient and accurate credit evaluation system from the following three aspects:

(1) Based on the rich customer basic information, financial transaction data of the enterprise's traditional database, and combined with customer credit data obtained from social media and Internet financial platforms, a complete customer credit data platform can be built.
(2) Using big data technology, we can combine the professional and quantitative credit model of financial enterprises, and evaluate the repayment ability and willingness based on the credit accumulated from the purchase, sales, payment and settlement, logistics, and other transactions of the Internet. We can also apply industry standards to restore real business conditions, analyze massive customer credit data, and establish a complete credit evaluation model.
(3) By applying big data technology, we can carry out distributed computing deployment of credit model, to achieve rapid response, efficient evaluation, and rapid lending, and to realize batch issuance of small and microloans and credit products for small and micro enterprises.

For example, for farmers' loans, Fujian Agricultural Bank currently allows farmers to loan with their own credit, which can proceed the loan quickly and achieve sustainable business operation of inclusive finance. Fujian's characteristic agriculture has obvious advantages but is limited by the hills and landforms, resulting in small agricultural planting and breeding scales. Consequently, the microloans for farmers have many difficulties, such as wide coverage, scattered objects, a large number of transactions, small amount, and large workload.

To address this, Fujian Agricultural Bank has innovatively launched the Internet small-amount farmer loan product "Fast Famer Loan" nationwide. Relying on the Internet and big data technology, it combines financial credit information, government affairs information, agricultural product market transaction records, credit village records, and other information, scientifically designing the credit extension model to realize the scale, standardization, and automation of farmers' loans.

In June 2018, the province's Agricultural Bank's "Fast Farmer Loan" balance was 4.66 billion RMB, supporting 50,000 farmers. Using modern technical means for traditional credit business has achieved "fast, accurate and favorable", and has effectively solved the "difficulty in guarantee, difficulty in financing, and unaffordable financing" problems of farmers.

High-frequency trading and algorithmic trading

Let's take high-frequency trading as an example. In order to gain profits, traders take advantage of hardware equipment and trading programs to quickly obtain, analyze, generate, and send trading orders, and buy and sell frequently in a short period of time. Today's high-frequency trading mainly adopts "strategic sequential trading", that is, by analyzing financial big data to identify the footprints left by specific market participants.

In this process, the full combination of big data and artificial intelligence technology can achieve a multiplier effect. A study by the Research and Development Department of China Securities Investment Corporation pointed out that quantitative funds based on artificial intelligence technology in China and other countries performed well. In foreign countries, Virtu Financial, a high-frequency program trading, only 1 out of 1,238 trading days was in the red; hedge fund Cerebellum hasn't posted a monthly loss since using artificial intelligence in 2009; the first Artificial intelligence-driven fund Rebellion had successfully predicted the 2008 stock market decline. In China, the quantitative trading driven by practical artificial intelligence is still in the initial stage.

Public opinion analysis of products and services

With the popularization and development of the Internet, financial enterprises not only expand more and more businesses online, but customers also increasingly choose to speak up through various network channels. Some negative public opinions of financial enterprises are quickly spread online, which may bring huge risks to the financial industry and even to the economy.

Financial institutions need to use public opinion collection and analysis technology to capture information related to financial institutions and products from social channels through big data crawler technology, and use natural language processing technology and data mining algorithms to perform word segmentation, clustering, feature extraction, and association. They then find out the market attention, positive and negative evaluations of financial companies and their products, as well as the user reputation of various businesses, especially the timely tracking and early warning of negative market public opinion, which can help companies detect and resolve crises in time. At the same time, financial enterprises can also choose to pay attention to the positive and negative information of competitors in the same industry, taking it as a reference for their own business optimization, to avoid missing opportunities.

Customer risk control

The risk control of traditional finance is mainly based on the credit data of the central bank and the ecological data in the banking system and is completed by manual review. Under the circumstance that the domestic credit information service system is not perfect, the core of Internet amount risk control lies in relying on the big data obtained from the Internet. Companies such as BAT have plenty of user information that can be used to predict the risks of microloans more comprehensively.

In the application scenario of enterprise data, the most commonly used models are supervised learning and unsupervised learning. A natural and typical application in the financial industry is the credit evaluation of borrowers with risk control. Therefore, Internet financial enterprises rely on the Internet to obtain users' online consumption behavior data, communication data, credit card data, third-party credit data, and other rich and comprehensive data, and use machine learning to build big data risk control systems for the Internet financial enterprises. Meanwhile, through the intelligent customer acquisition model, the demand analysis and risk estimation of the customer group can be carried out. Based on the judgment of the user's needs, credit, and risk levels, as well as the match with the product, accurate portraits are made to provide "thousands of people and thousands of faces" services. This could provide personalized services and reduce customer acquisition costs.

In addition to the credit review before lending, Internet finance companies can also use machine learning to monitor the borrower's repayment ability in real time during the lending process. In case someone may not be able to repay the loan in the future, it could deal with it in a timely manner, thereby reducing losses caused by bad debts.

Targeted and Personalized Poverty Alleviation

Targeted poverty alleviation mainly contrasts with extensive poverty alleviation. In the past, in the process of implementing extensive poverty alleviation, it was faced with difficulties such as accurately identifying the poor, assisting the poor, collecting data, and monitoring the poverty alleviation results.

In contrast, targeted poverty alleviation achieves the optimal allocation of resources through precise identification, precise assistance, precise management, and precise assessment. In the process of precise poverty alleviation, it is necessary to use the power of big data to find out the poor people who really need help, find out the reasons for poverty, put the assistance measures in place, and conduct poverty alleviation and achievement evaluation.

Set targeted poverty alleviation goals based on big data analysis

In recent years, governments at all levels continued to strengthen the support for poverty alleviation, and the financial input is increasing day by day. However, in the specific operation process, how to determine the project, how to use the funds, and how to do the work are the key tasks that poverty alleviation workers need to plan in the overall level.

In this process, big data plays an increasingly important role. By building a big data precision assistance platform, and conducting real-time dynamic monitoring, analysis, and assessment of poverty alleviation data, poverty alleviation workers can not only identify the main body, focus, and key of poverty alleviation but also evaluate whether the poverty alleviation project is truly scientific, reasonable, and accurate. In this way, poverty alleviation funds and valuable resources could be accurately allocated to the real poor households.

Accurate targeting of poverty alleviation based on big data analysis

Poverty alleviation must first identify the poverty. The primary difficulty is "whom to support". Due to the multi-dimensional, complex, and dynamic characteristics of poverty, it is difficult to accurately identify the "true poverty". In order to solve this problem, more and more government poverty alleviation departments have begun to share, integrate, and record the collected data of poverty alleviation objects through the construction and application of big data precision poverty alleviation platforms and service systems.

In the precise poverty alleviation system established with big data technology, comprehensive information on poverty alleviation objects can be found and presented (including macro information, such as the number of poor people and regional distribution, as well as specific information such as the family situation and education level of the poverty alleviation objects). Then, we can conduct multi-party analysis, comparison, and needs assessment to make it a three-dimensional object, and screen out the targeted person that truly needs help. At the same time, by further integrating the data of civil affairs, finance, disabled persons federation, social security, real estate bureau, and other departments, the "information island" between regions and departments will be broken, and the scattered and fragmented information will be networked and integrated.

Taking Guiyang City as an example, it has integrated various industry data and built a unified poverty alleviation database by launching a big data targeted assistance platform. By July 2017, the database has accumulated 6.913 million pieces of data, including nearly 520,000 pieces of poverty alleviation data for low-income groups. On this basis, through the establishment of a "poor population identification model", and combining the business data of multiple relevant departments with the traditional identification model of "two public announcements and one announcement" for the poor, the model could quickly and accurately identify low-income groups and families in difficulty.

Accurately help the poor based on big data analysis

After identifying the specific assistance person, big data technology can be used to dig out the true causes of poverty and the need for "real poverty" by analyzing the poverty manifestations and causes in multidimensional states. This will help policymakers implement targeted poverty alleviation measures and better achieve targeted poverty alleviation that matches supply and demand.

For example, in the targeted poverty alleviation system built by Chongqing Mobile, it, on one hand, provides poverty alleviation cadres with more than 10 functions such as news announcements, poverty alleviation policies, village check-in, work account, case sharing, and poor household inquiry, which enables them to better understand the targets of poverty alleviation; on the other hand, it also opens an informatization window for poor households to help them obtain policies that benefit in a timely manner.

At the same time, in specific poverty alleviation work, by tracking the implementation of poverty alleviation policies and the use of funds, it is possible to carry out dynamic early warning and precise management of poverty alleviation work on the basis of big data technical analysis and processing. Based on this, tracking and early warning and decision-making optimization can be carried out to achieve precise management that can ensure the accurate placement of poverty alleviation policies and the effective use of poverty alleviation funds.

Evaluation of poverty alleviation effect based on big data analysis

How to accurately evaluate the effectiveness of targeted poverty alleviation requires that the evaluation content be concreted into various indicators. In this process, a set of poverty alleviation assessment index systems and digital assessment systems can be embedded in the poverty alleviation information management system. Then, integrate diversified data (such as the decline rate of the number of poor people in a certain area as a percentage of the total population in that area and statistics on the types of social services the poor people receive), and using big data technology to extract, organize, analyze, and model the relevant data, to implement dynamic monitoring of the assessment results and quantitative assessment of the whole process.

Based on that, the management department will provide information services and references for the next stage of poverty alleviation goals, methods, priorities, capital investment, etc., and thus ensure the accurate implementation of assistance policies, and could effectively help the poor out of poverty.

In the fifth issue of *China Poverty Alleviation* in 2017, 12 typical cases of targeted poverty alleviation collected by the State Council Poverty Alleviation Office were recorded. Among them, Shibadong Village is a Miao village in Hunan Province. In 2013, there were 225 households with 939 people, and the per capita arable land was only 0.8 mu (1 mu ≈ 0.067 hectare). The per capita net income is only 41% of the county's average level (4,903 RMB). There are 136 registered poor households and 542 poor people, and the poverty incidence rate is as high as 57.7%.

In 2013, General Secretary Xi Jinping visited Shibadong Village, where he put forward the important idea of "targeted poverty alleviation" for the first time, and made important instructions of "seeking truth from facts, acting according to local conditions, conducting classified guidance and achieving targeted poverty alleviation". He also demanded that Shibadong Village "cannot be specialized, but can never stay unchanged". Since then, Shibadong Village has entered the fast lane of targeted poverty alleviation.

In the past few years, Shibadong Village has mainly done three things to implement targeted poverty alleviation: First, elect the first secretary and build a strong village party branch. The village team quickly sorted out development ideas, based on big data analysis, accurately identified poor households, formulated poverty alleviation measures one by one, and demonstrated development projects one by one. The second is to raise funds of 27.13 million RMB from various parties to improve infrastructure and public services. The road into the village was widened and strengthened by 3 kilometers, the walkway and guardrail were built in the village, the main water supply pipeline was repaired to solve villagers' production and living water shortage, the renovation of houses, kitchens, toilets, baths, and neighborhoods was completed, the access of radio and television to every household was realized, two primary schools were repaired and renovated, and two village clinics were newly built. The third is to adapt measures to local conditions and develop characteristic industries precisely. To develop characteristic planting industry, they focused on the development of flue-cured tobacco, kiwi fruit, wild vegetables, winter peach, camellia, and other planting. To develop characteristic breeding industry, one should focus on the development of cattle, pigs, and paddy fish farming in western Hunan. To develop characteristic processing industry, they should focus on the development of Miao embroidery and brocade. To develop characteristic rural tourism, eight farmhouses were opened relying on natural landscapes and folk customs.

In just three years, Shibadong Village has achieved significant changes. The per capita net income of the whole village increased from 1,668 RMB in 2013 to 8,313 RMB in 2016, with an average annual increase of 2,215 RMB and an average annual increase of 130%. All poor households in Shibadong Village have been lifted out of poverty, and the village has shaken off poverty.

Chapter 8

Big Data Serves People's Livelihood and Welfare

The data are unlimited yet the value is visible. Whether it is data collection, cleaning, or post-processing and analysis, its purpose is to serve the people. In various livelihood issues, the application of big data is directly related to the vital interests of the people. It is necessary to make full use of big data's technical power, eliminate "data islands", support social security, intelligent transportation, intelligent environmental protection, intelligent medical treatment, education applications, etc., and better serve people's livelihood and well-being.

Law and Order Under the Protection of Big Data

As *The Book of Lord Shang: The Elimination of Strength* said: *A strong country knows thirteen figures: the number of granaries within its borders, the number of able-bodied men and of women, the number of old and of weak people, the number of officials and of officers, the number of those making a livelihood by talking, the number of useful people, the number of horses and of oxen, and the quantity of fodder and of straw. If he who wishes to make his country strong, does not know these thirteen figures, though his geographical position may be favourable and the population numerous, his state will become weaker and weaker, until it is dismembered.* In layman's terms, national governance is inseparable from the granary, population, cattle and horses, and other customs and properties. To strengthen a country and govern, a country must first be based on the national conditions of "the thirteen numbers of the country". This shows Shang Yang was the earliest pioneer in governing the country with data.

The prevention and control of social safety is a key topic to be addressed in the governance of the country. Currently, with the support of big data, social safety is gradually developing in the direction of informatization and is quietly changing the ecology of national governance. In the era of big data, the public safety agency comprehensively collects and integrates massive data, process, analyze, and deeply mine the data to discover the inherent laws of the data, providing strong support for preventing and combating crimes. Promoting the construction of public security informatization with big data is an important way to improve the efficiency of public security work, and it is also an advanced form of public security informatization application.

Integrate data resources for unified analysis and judgment

In the past, the police often needed to track and find clues for a long time to catch the recidivist thieves. Now, in an environment where big data is widely used, every move of a thief is very likely to leave a digital trace. For example, through the payment platform, it is quite easy to notice a suspicious person who took 30 buses every day. This is the power of data resources that cannot be ignored in public safety prevention and control. Under the nourishment of big data technology, the innovation of social security prevention and control systems with data governance is just at the right time.

In the process of public safety management, government functional departments can collect big data such as floating personnel, temporary residents, visitors, vehicles, and sales information of key elements, forming a data network covering the whole society. With the help of the big data management and analysis platform, they can collect, organize, archive, analyze and forecast massive data, dig out the inherent, inevitable causal relationship behind all kinds of data from complex data, and find hidden laws. Based on this, real-time tracking, real-time research and judgment, and real-time disposal could be achieved, and the three major problems of difficulty in prevention and management in advance, difficulty in first-time detection, and difficulty in detecting cases after they occur could be eliminated.

American company PredPol has launched a criminal activity prediction platform. Its main interface is a city map just like Baidu Map, which is similar to our police geography platform. Based on the statistics of past criminal activities in a certain area and with the help of algorithms, it calculates the probability, type, and the most likely time period of the crime, and marks the key areas that need high vigilance with red boxes. The on-duty policeman can check online through computers, mobile phones, etc. Thereby, through targeted patrols and prevention, the intelligent analysis and prediction of crime information could be realized.

In China, there are also many urban service applications. For example, "My Nanjing" is a city-level public service mobile application that integrates various life information, including relevant service resources and authoritative information of relevant government departments and public institutions. It provides the public with information closely related to individuals, and also provides information services on life, medical care, transportation, tourism, convenience, government affairs, information, and affairs.

"My Nanjing" is a real-name authentication mobile app. The personal information submitted needs to be verified with the back-end database before the successful registration. After successful registration, one can enter the "my channel" to view personal information services such as provident fund, social security, tax payment, personal credit, driver's license, vehicle violations, and living information services such as the usage of gas fees easily and conveniently. All of the information demonstrates the people-oriented service concept.

The users do not need to register for the "city channel", which provides various service information related to the city, including transportation, medical services, tourism services, government services, community services, and public welfare services. It also provides urban information services such as local weather forecast and real-time air quality index. In the "Nanjing Information" channel, you can check the information of Nanjing Express, such as the development and reform news, people's livelihood information, and convenience services, view the public documents of the government, such as municipal government documents, municipal government office documents, policy interpretation, public announcements, government announcements, and information disclosure annual reports. You can also open online services and interactive exchanges for enterprises and individuals, covering social security, entrepreneurship and employment, audit information, market supervision, quality supervision, social organizations, and other key areas.

At present, the "My Nanjing" mobile terminal software has Android and iOS versions, which can be downloaded from the mobile app store. Another feature is that it provides rich and flexible intelligent voice services. The user speaks the content to be searched to the mobile phone, and then the required content can be obtained immediately through intelligent voice analysis. The search response is fast and the content feedback is accurate, allowing more end users to get the convenient operation experience brought by the city smart portal. Meanwhile, it also provides 50 GB free space of personal cloud box for each registered real-name system user. Users can store their own files, photo albums, videos, etc. in a private cloud box, and can also back up the contacts and text messages from the mobile phone into the cloud box, thus enjoying the convenience of obtaining files and data in the cloud box at anytime and anywhere.

In addition, with the support of Video cloud video technology, users can also view the real-time traffic conditions of Nanjing's main roads through the "road condition big data" function. Before the app was launched, the Smart Nanjing Center had applied the cVideo cloud video surveillance system to realize the connection with the existing video surveillance platforms such as the Traffic Management Bureau, the Traffic Bureau, the Public Security Bureau, the urban high-point surveillance, and the "320" project of road image surveillance, which has provided powerful video messages for security, transportation, environmental monitoring, emergency command, and other fields (Figure 8.1).

The "My Nanjing" application terminal will further establish social sports information services (integrating resources such as stadiums, clubs, members, and coaches, and building a sports social platform integrating fitness, making friends, and shopping), smart medical information service (integrating the roles of patients, medical staff, and medical institutions, to establish a health security system that is closest to the needs of citizens), government affairs public information services, intelligent transportation planning, online payment platform docking, location services, etc.

Integrate departmental strength for unified scheduling and processing

In the construction of big data in public security control, it is necessary to integrate network platforms and departmental forces to avoid "Isolated

Figure 8.1   "My Nanjing" City Intelligent Portal Monitoring and Analysis Platform

Data Island" and provide a unified big data application environment with centralized resources, centralized management, centralized monitoring and supporting implementation, so as to provide support, service, and guarantee for the practical application of overall social security. Therefore, many cities are currently committed to opening a comprehensive management information platform, realizing the full coverage of the network information sharing platform, and fully integrating the data resources and information of the public security, justice, civil affairs, health planning, forestry, land, and other departments, and finally realize the interoperability, interconnection, sharing, unified linkage, and unified command with departments and industries, thereby improving emergency response capabilities.

In 2014, Jiangsu Province developed a comprehensive management information system, and achieved the five-level coverage of provinces, cities, counties, towns, and villages, including 10 basic business modules and 5 auxiliary modules. In October 2015, through unified deployment, a comprehensive management information system was further built, covering 12 streets and 106 communities in the district. At present, Nanjing has been connected to the provincial comprehensive management information system, focusing on building a large-scale comprehensive management system in Qinhuai District, building a complete set of work systems and workflows, and providing a unified data collection channel, information display channel, information processing center, and command and coordination center.

Based on the comprehensive management information system of Jiangsu Province, the comprehensive management of Qinhuai District carries out business function expansion and business in-depth application that mainly focuses on the following five construction priorities: First, the key departments realize the integration of resources and realize the sharing of data between departments. Second is to achieve the management and control of key areas such as video surveillance of commercial streets, key rivers, communities, and crowded places through intelligent video technology. Third is to achieve the management and control of key personnel, based on existing data and portrait recognition technology, to realize real-time control of the person's status. Forth is to prevent key sensitive issues. Fifth is to address the key concerns of citizens, such as parking difficulties, air pollution, black and odorous rivers, public security, and other issues, and gradually realize multi-faceted urban refined management.

Note that in the process of comprehensive urban management and safety construction, real-time panoramic monitoring of road traffic and pedestrians has increasingly become the prerequisite and foundation. Smart Street Light Companion, released at the 2018 Digital Economy Conference, provides a solution. Through six cameras, the Smart Street Light Companion can monitor the road and pedestrians at 360° with no dead angle, and display it in real time on its platform to provide services for managers to conduct real-time inspections.

In addition, the Smart Street Light Companion integrates a variety of sensors that can dynamically monitor changes in air pollution such as $PM_{2.5}$ and PM_{10} in real time, which can dynamically analyze the development process of urban pollution, and realize the location and prevention of pollution sources. Moreover, it can also provide various functions such as emergency charging, traffic monitoring, and convenient information interaction, which is helpful for route inquiry and emergency response, and can also help find lost people and criminal suspects.

Undoubtedly, with the continuous emergence of applications similar to Smart Street Light Companions, the application scenarios in the construction of safe cities in the future will be increasingly enriched, which will further help upgrade urban management including public safety, urban management, transportation, and environmental protection, strengthen public security management and control, fight against illegal and criminal activities, track environmental changes, and make urban management safer, more orderly, and more convenient.

Smart Transportation with Artificial Intelligence

To address the disadvantages of weak core technical capabilities, insufficient resource integration, and difficulty in exerting the overall system functions in traditional transportation applications, intelligent transportation, as the development direction of future transportation, uses electronic police, bayonet and traffic signal control lights, etc., with the help of information technology, electronic sensing technology, and data communication technology to establish a real-time, accurate, and efficient integrated transportation management system.

In the field of transportation, massive traffic data is generated from the operation monitoring and service of various types of traffic, various traffic and meteorological monitoring data of expressways and arterial roads, and GPS data of buses, taxis, and passenger vehicles. The amount of data is large and of many types, also which has vastly increased from the terabyte level to the petabyte level. In Guangzhou, more than 1.2 billion new urban traffic operation data records, with a volume of 150–300 GB, are generated every day.

As professor Daniel Esty of Yale Law School pointed out: "Based on a data-driven approach to decision-making, governments will be more efficient, more open, and more accountable, and it will be guided by evidence-based facts". In the era of big data, big data technology promotes intelligent transportation towards comprehensive informatization. Through data integration and mining applications, it provides an important reference for the government to promote the implementation of intelligent transportation.

Intelligent road condition analysis to improve road efficiency

Through the comprehensive application of communication technology, sensing technology, control technology, information technology, artificial intelligence technology, etc., a comprehensive real-time highway monitoring system can be realized. It quickly collects diverse data such as vehicles and pedestrians through various functional systems, provides applications such as data query, analysis, and processing, and provides support for traffic management and traffic information services. For example, relying on big data analysis and processing technology, the traffic management department can know the accident, construction, and real-time road conditions in real time. It can also obtain weather conditions such as rain, snow, and heavy fog, and publish them on the corresponding mobile travel terminals to facilitate the planning of travel time, optimal routes, and transportation methods. For example, by building a road traffic index through real-time acquiring big data such as congested road sections and time, it is possible to quickly find out the rules and make predictions, which can consequently provide a basis for urban traffic management and maximize the road traffic efficiency.

For example, Alibaba is building a super artificial intelligence for city governance, the city brain. By adopting Alibaba Cloud ET artificial intelligence technology, it can realize the global real-time analysis of the city and automatically allocate public resources. Currently, through various types of data perception of traffic situations and optimization of signal light timing, Alibaba Cloud ET City Brain has helped reduce the travel time in the pilot area by 15.3%, saving an average of 4.6 minutes per vehicle for major elevated travel time.

Intelligent behavior analysis facilitates illegal monitoring

In road monitoring, by combining artificial intelligence technologies such as image intelligent analysis algorithms, and with the help of intelligent transportation platforms, it could accurately identify the vehicle color, model, license plate, and other attributes, and can also monitor and analyze the driver's behavior (including whether to wear a seat belt and whether to answer a phone call). This helps check the source of accidents, and provides reference and technical support for traffic police law enforcement.

For example, the video recognition algorithm of Alibaba Cloud ET enables the urban brain to perceive the running trajectories of vehicles on complex roads, with an accuracy rate of over 99%. In addition, the City Brain integrates AutoNavi, microwave, and video data of traffic police to perceive traffic events, including congestion, illegal parking, and accidents, and trigger mechanisms for intelligent processing. In the main urban area, the average number of daily event alarms in the urban brain reaches more than 500 times, with an accuracy rate of 92%, which has greatly improved the directionality of law enforcement.

Intelligent research and analysis to provide decision-making basis

For a long time, most of the various traffic data we collected simply stayed in the "viewing" stage, and mostly only provide evidence for post-event analysis. Now, through cloud computing, big data, artificial intelligence, and other technologies, the establishment of an intelligent traffic data research and judgment system connected to the urban bayonet system can provide a basis for in-depth analysis after the event. Also, more importantly, it can cross-regionally, cross-departmentally integrate the data and resources, and provide early warning and prediction of traffic behaviors or events, thereby providing the decision-making basis for the optimization of corresponding departments such as public safety, criminal investigation, and traffic management, and helping to respond quickly during events.

For example, in response to the problem of traffic light control in the city, Dr. Liu Peng of Tsinghua University led the research and development team using the idea of AlphaGo. Based on the traffic flow conditions in Nanjing, the composition of data and algorithms is fully applied to dynamically allocate traffic lights, which will greatly alleviate traffic congestion and improve traffic efficiency in the future.

In addition, on the Didi intelligent transportation cloud platform, through integrating sensor data, static road data, road event data, as well as Didi's traffic data, driver data, GPS trajectory data, and capacity data, they achieved a regional heat map, traffic volume data analysis, urban capacity analysis, urban traffic forecast, etc., which provide important references for urban public travel.

Big Data Environmental Protection for Grid Monitoring

In environmental protection and governance, in order to maximize the role of ecological and environmental protection big data, it is urgent to real-time collect and update the data through widely deployed sensing equipment to create an "environmental protection network" to realize grid monitoring, and then quickly determine the pollution source and carry out traceability management.

Big data and environmental monitoring

The following lists the cases of big data and air monitoring as well as big data and earthquake monitoring, from different angles.

Big data and air monitoring

For environmental monitoring and governance, on the one hand, it is important to apply big data technology to clarify WHO is polluting, so as to facilitate prevention and control from the source; on the other hand, after identifying the source of pollution, the management department needs to strictly enforce the law to achieve efficient supervision and governance. In recent years, with the further implementation of the "Environmental Protection Law of the People's Republic of China" and the ever-increasing law enforcement efforts, serious law enforcement is no longer the crux of the matter, but pollution traceability has become the critical problem to be solved.

In environmental protection, making good use of big data is an obligatory course. Among them, environmental monitoring is the most closely related link with big data, which is also the starting point for the application of big data technology. Extensive collection of atmospheric data, meteorological data, water quality data, and other environmental protection data can effectively "speak with data" in environmental protection.

Environmental pollution encounters problems such as large discharge load and serious compound air pollution. In order to qualitatively and quantitatively monitor and analyze pollutant factors, a large number of monitoring equipment and sensing terminals need to be applied. However, with the continuous expansion of the deployment scale, the monitoring cost will rise sharply, which will inevitably become a major bottleneck restricting the application.

For the typical problem of $PM_{2.5}$, the environmental protection department generally manages and controls through the construction of atmospheric monitoring stations. Although the measured value is accurate, the cost is high, and thus it is difficult to deploy monitoring stations on a large scale. Therefore, the limited number of monitoring points at all levels, such as national control and provincial control, has become a major challenge for traditional monitoring methods. Since 2011, Yunchuang Big Data began to pilot large-scale grid deployment of $PM_{2.5}$ cloud monitoring nodes, using high-precision imported sensors to collect data, and upload it to the background data processing platform for analysis and display, dynamic tracking, source of pollution, and pollution process locating. This has changed our passive situation in atmospheric monitoring. At present, the number of $PM_{2.5}$ cloud monitoring nodes deployed has reached 5,309 nationwide.

Since then, on the Kunshan Qiandeng Environmental Protection Air Online Monitoring Platform, through the 80 monitoring points set up in the park (each node can monitor four main pollution indicators, and the monitoring indicators of each node are different), it achieved real-time monitoring of hydrogen chloride, sulfur dioxide, epichlorohydrin, all indoor organic gaseous substances, methanol, acetone, ammonia, and other harmful substances in enterprises in the industrial zone. So far, the platform has serviced for more than 5 years. Under the effective guidance and management, the air pollution in the industrial park and Kunshan has been well controlled.

At the same time, on the Environmental Cloud, that is, the Environmental Big Data Open Platform (www.envicloud.cn), through mining and analysis of historical environmental data, the correlation between certain environmental data can be found, such as the weather changes before and after earthquakes and the influence of meteorological conditions on the diffusion of atmospheric pollutants. A better environmental data model can be established by summarizing the data rules, thereby improving the accuracy of environmental data prediction.

Big data and earthquake monitoring

In addition to air monitoring, earthquake monitoring is also a top priority. Traditional methods have higher requirements on capital and manpower for the construction of intensive network, while high-tech MEMS sensor technology, new sensor technology, mobile Internet, and sensor network technology bring low-cost, reliable, and intelligent new seismic observation equipment for earthquake early warning and intensive seismic observation network. These devices not only meet the needs of earthquake early warning but also will continue to develop as the technology continues to develop to further matches the needs of real-time seismic observation, and its cost is 1/20 of that of traditional seismographs.

At present, the China Earthquake Administration is testing the use of eCat indoor environmental detectors with earthquake early warning functions for grid monitoring layout. Figures 8.2–8.4 are MEMS earthquake early warning station equipment produced by Sichuan Chengdu High-tech Disaster Mitigation Research Institute and other units, as well as environmental cat indoor environmental detectors.

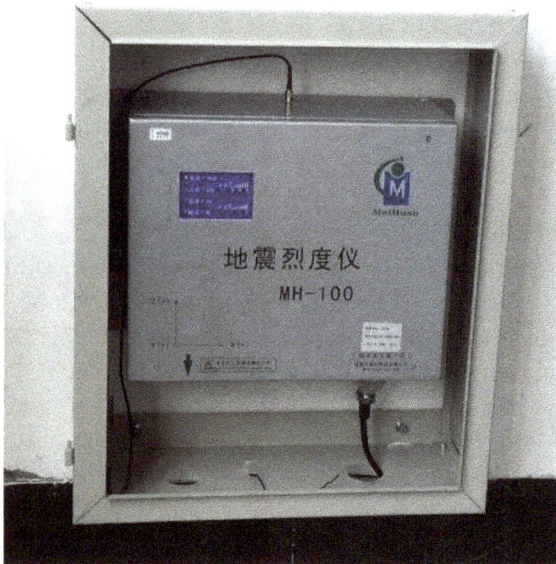

Figure 8.2   Production of Sichuan Chengdu High-Tech Disaster Reduction Research Institute: MEMS Earthquake Early Warning Station Equipment

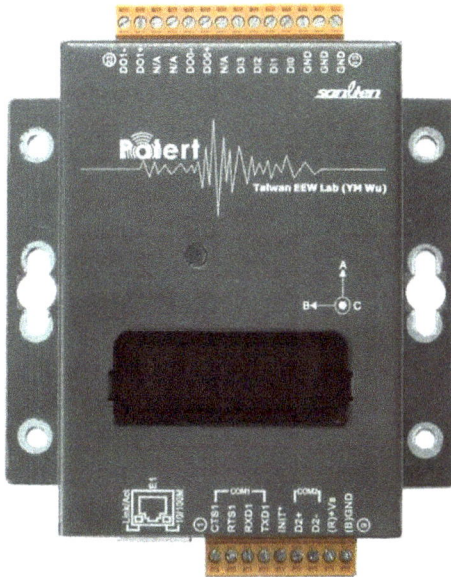

Figure 8.3   MEMS Earthquake Early Warning Station Equipment

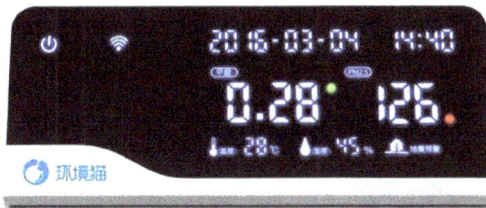

Figure 8.4   Environmental Cat Indoor Environmental Detector

Big data and environmental management

In the process of environmental management, through extensive monitoring of various environmental data, we can conduct real-time simulation of pollutant distribution and its evolution rules through numerical model source tracing and process analysis technology. Based on the simulation results, analysis of the evolution of pollutants in different heavy pollution processes, study of the contribution of different sources to pollutant concentrations, and the analysis and tracking of the concentration, source, and destination of various major pollutants can be obtained. This could provide an important reference for environmental management.

For example, based on extensive and accurate data collection, the application of big data technology and deep recurrent neural network (DRNN), etc., can provide strong technical support for real-time monitoring, pollution traceability, pollution process evolution, refined forecasting, precise supervision, and publishing services, establish a construction system integrating environmental monitoring, early warning and traceability, and play an important role in the realization of grid-based, refined and scientific environmental supervision by environmental protection departments (Figure 8.5).

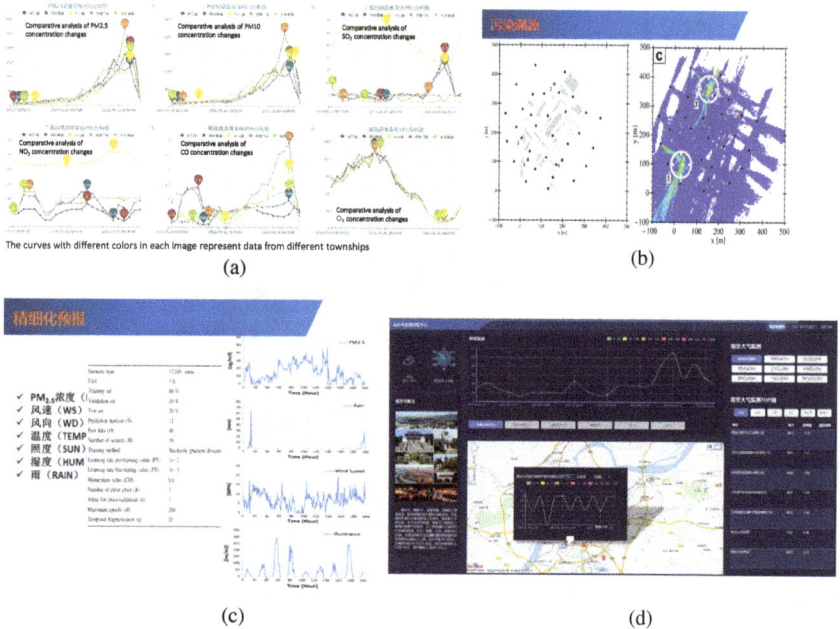

Figure 8.5   Construction System Integrating Environmental Monitoring, Early Warning, and Traceability. (a) Real-time monitoring (b) Pollution traceability (c) Fine monitoring (d) Monitoring and early warning

In terms of river management and protection, it is a complex project involving upstream and downstream, left and right banks, different administrative regions, and industries. In recent years, some regions have actively explored the river chief system, with the party and government leaders serving as river chiefs, implementing local main responsibilities in accordance with laws and regulations, and coordinating and integrating all forces to promote water resources protection, water coastline management, water pollution prevention and control, water environment management, and other work.

Based on this, the Smart River Chief Project, guided by the "Opinions on the Comprehensive Implementation of the River Chief System" of Jiangsu Province, has developed a "Smart River Chief" water quality monitoring and early warning system. The system is city, town (street) hierarchical management. It integrates various existing basic data, monitoring data and video, uses the transmission network to quickly converge to the monitoring and early warning system, and provides inquiries, reports, and management systems at different levels, different dimensions, and different carriers for leaders at all levels, river chiefs, staff, and the public (Figures 8.6 and 8.7).

Specifically, through grid monitoring, data visualization, big data mining and analysis methods, real-time monitoring and early warning of river and lake water quality can be realized, and river chiefs at all levels

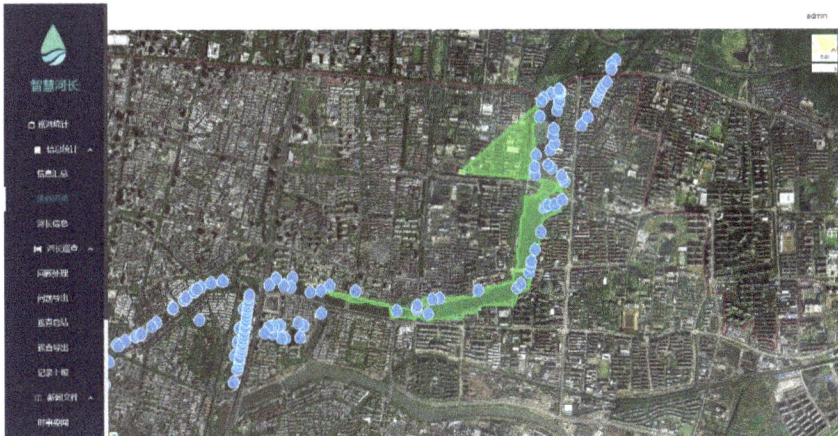

Figure 8.6   Homepage of Smart River Chief Application

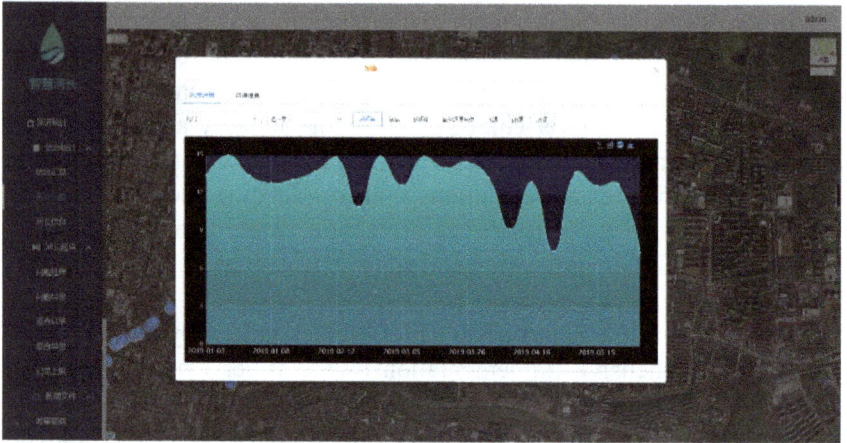

Figure 8.7   Query of River Water Quality

can understand the basic information of rivers under their jurisdiction, including the distribution of key monitoring enterprises according to the classification of direct discharge and sewage treatment plants, the distribution of sewage treatment plants, the status report of water quality, the distribution of river length signs, and the sand mining, sand mining and other river water environment treatment work plans and progress. Through the establishment of river archives and river management strategies, "one river, one file" and "one river, one policy" are formed to assist river chiefs in various regions to comprehensively promote the environmental protection and restoration of rivers and lakes, thereby improving the water quality and water environment of rivers and lakes, and promoting the coordinated development of economy, society, and ecological environment.

Smart Healthcare Supported by Big Data and Artificial Intelligence

Big data has now been widely used in the medical field. In the fields of medicine and health care, such as disease treatment, clinical research, new drug research and development, and early warning of infectious diseases, all of them involve data collection, management, and needs analysis. The construction of smart medical care aims to rely on big data and artificial intelligence technology to further reduce the workload of doctors and alleviate the current situation of insufficient and uneven medical resources.

Integrating medical resources with a three-dimensional database

In the combined application of big data and medicine, through making full use of the resources of medical big data, the platform of the medical and health network and the database of health records are established, which provides great convenience for the diagnosis and treatment of diseases and medical research. When patients seek medical treatment, the electronic medical records, health files, and other personal information provide comprehensive health index information of patients, which helps doctors to diagnose and treat diseases quickly and accurately, and further improves the quality of diagnosis and treatment. Meanwhile, it can provide citizens with convenient and beneficial applications such as online appointment and triage, sharing and mutual recognition of test results, and remote settlement of medical insurance online.

In addition, on the basis of the health record database, by integrating medical big data resources, a support system such as clinical decision-making, disease diagnosis, and drug research and development will be constructed to further expand applications such as public health monitoring and evaluation, and early warning of infectious disease epidemics, which help medical staff to further achieve "prediction, early warning, and prevention" on the basis of "calculating data, knowing data, and using data".

The collection, mining, and utilization of medical and health big data such as electronic medical records and residents' health records have improved the intelligence of clinical decision-making in medical institutions and the remote patient monitoring accuracy, improved the efficiency of public health and public health monitoring in health departments, shortened the research and development cycle of medical drugs in scientific research institutions, and provided an effective decision-making basis for the whole society to prevent and control large-scale epidemics, optimize the allocation of medical resources, and protect people's health.

AI diagnosis based on medical big data

Nowadays, the early identification and diagnosis of diseases with artificial intelligence technology are not new. A study by Harvard Medical School and the Massachusetts Institute of Technology using deep learning algorithms on breast cancer showed that a pathologist's accuracy rate was 73.3% in determining the tumor score, while AI's accuracy rate could be

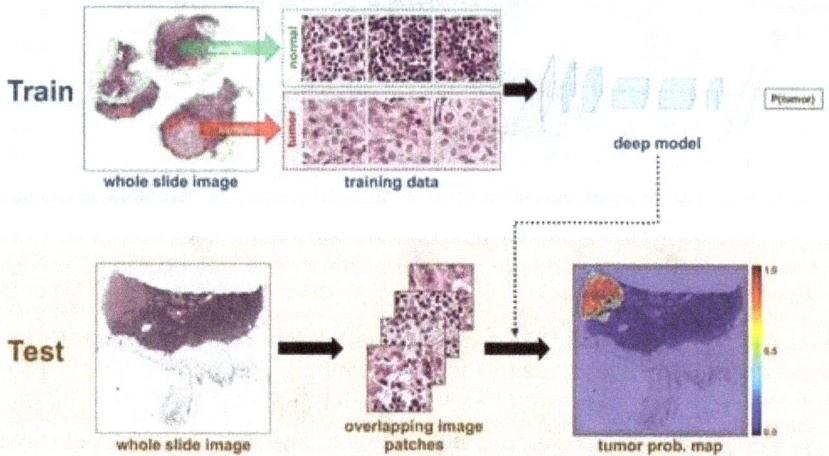

Figure 8.8   AI Cancer Cell Detection Process

as high as 96.6%. If AI prediction can be used as a medical aid in the early stage to provide an important reference for doctors in later diagnoses, the prediction accuracy rate is expected to reach 99.5%.

For example, as shown in Figure 8.8, in the process of breast cancer detection and diagnosis, a large number of pathological slice images are divided into training set and test set, in which the training set data is used for deep learning training so that the network obtains the ability to identify diseased areas and healthy areas, and the test set is used to further judge the detection accuracy. Through continuous training and testing, the artificial intelligence system for cancer detection will gradually gain a higher ability to identify cancer, and then it can identify the lesion area in the form of a heat map.

Currently, based on the application of the deep learning all-in-one machine that can be widely used in image recognition, speech recognition, language translation, and other fields, with a maximum of 176 trillion single-precision computing power per second, through the AI training using pathological slices, the accuracy of predicting prostate cancer with deep learning has reached 99.38% (under binary classification) (Figure 8.9). It is developing from binary classification to multi-classification, and also from determining whether there is cancer to refining the classification of cancer malignancy and severity.

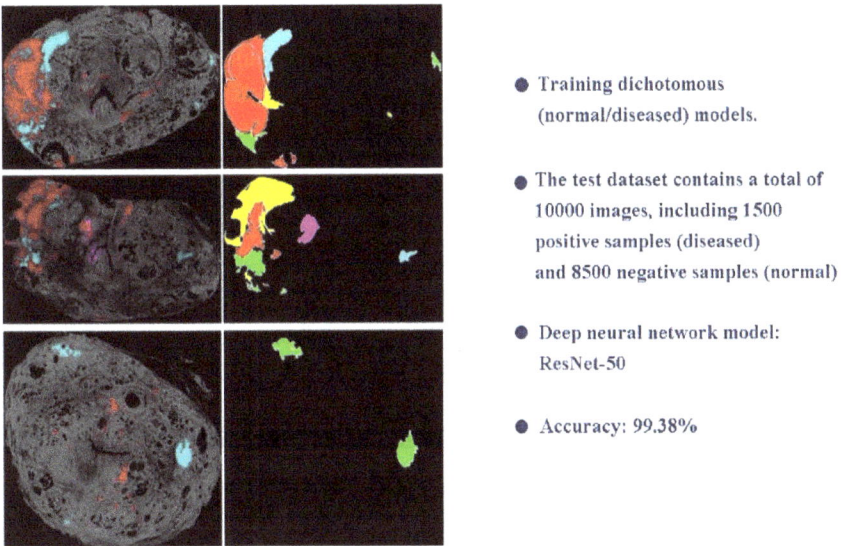

Figure 8.9   Intelligent Identification of Prostate Cancer

- Training dichotomous (normal/diseased) models.

- The test dataset contains a total of 10000 images, including 1500 positive samples (diseased) and 8500 negative samples (normal)

- Deep neural network model: ResNet-50

- Accuracy: 99.38%

Big Data and Education

The emergence of big data has changed the traditional education model and method, and can provide teachers and students with personalized teaching and learning plans, better promote personalized teaching and precision teaching, and help teach students in accordance with their aptitude, which could even directly promote the educational revolution.

"Deep Internet + artificial intelligence" calls for Education 3.0

Human society has entered an era of deep interconnection. Changes in the environment force education to make corresponding changes. The traditional book education and classroom education are challenged by changes. At the same time, the technological revolution of artificial intelligence is sweeping the world. We can call the stage from now on the fourth technological revolution.

Under the background of "deep interconnection + artificial intelligence", Education 3.0 emerged with the following six transformations as the most fundamental features.

(1) **From centralized learning to distributed learning:** Intensive classroom learning is no longer a necessary condition, and it is now possible to study online and take exams. For example, watching public courses with high-quality content leads to better learning results. The boom of large open online courses is sweeping the world.

(2) **From passive learning to active learning:** We need to develop a model based on big data and artificial intelligence technology, adopt intelligent teaching and anti-cheating online examinations, fully arouse students' interest, and allow students to change from passively listening in the classroom to actively learning through the Internet.

(3) **From participatory learning to immersive learning:** At present, virtual reality and augmented reality technologies have gradually matured. The teaching effect will be greatly improved if this intuitive and vivid interactive method is applied to teaching, since in this way students can transition from participatory learning to immersive learning.

(4) **From manually learning to intelligent learning:** Artificial intelligence can solve many educational problems. For example, students can interact with robots as long as there exists a teaching platform, and they can get advice from teachers only when encountering problems that cannot be solved by robots. This can save a lot of time and cost. There was a university in the United States that used a robot as a teaching assistant, yet, surprisingly, it was not uncovered for more than half a year.

(5) **From batch processing to personalization:** At present, teachers' teaching is still "one-to-many" batch processing, yet with the application of big data analysis technology and artificial intelligence, machine learning and continuously improved databases can be used to help students, and one can provide specific guidance according to the characteristics of individual students, enabling the transition from batch to personalization.

(6) **From learning knowledge to learning ability:** Many people complain that the knowledge they learned in college is not useful. In particular, for many computer science students, they even have to attend training courses for half a year after graduation before they can find a job. This is

because what they learn in school is knowledge instead of ability. Ability is only acquired when students can identify problems, solve problems, and quickly complete certain tasks and goals.

Big data + AI empowering education

At present, big data + AI is empowering all walks of life, and education is no exception. Intelligent technologies such as face recognition and speech recognition have begun to be used in Chinese, English, music, and other disciplines to provide education with more intelligent, personalized solutions (Figure 8.10).

For example, some teaching platforms can accurately locate various professional directions and integrate functions such as teaching, experimentation, training, and examination. On one hand, based on big data technology, they could establish a knowledge map, provide interactive teaching, and perform behavior data analysis. While dismantling the knowledge points for students, according to their feedback on the mastery of the knowledge points, they can check and fill in the gaps, and thus achieve personalized explanation and greatly improve the quality of teaching. On the other

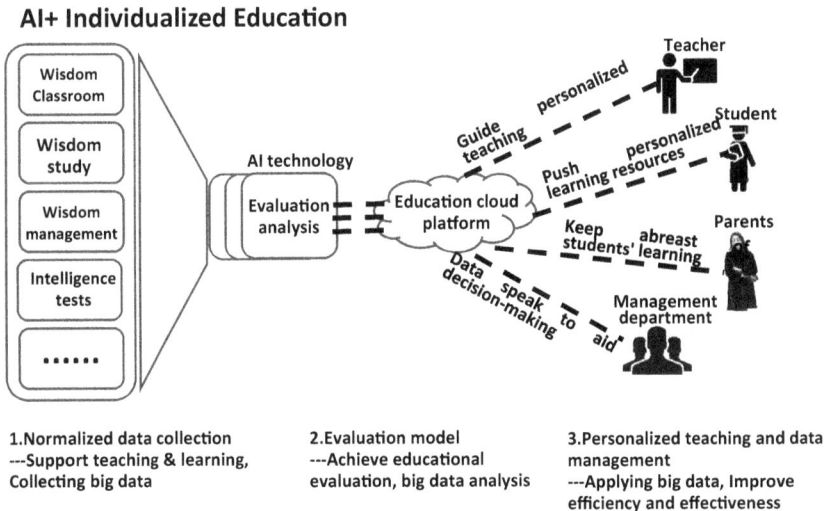

Figure 8.10   Big Data + AI Empowers Education

hand, with the support of artificial intelligence technology, the platform can answer various questions accurately and intelligently through the AI assistant, reducing the teaching intensity and improving the teaching effect.

It is obvious that the application of big data + AI in the field of education is becoming more and more extensive. If we look at the teaching process and implement it into teaching, learning, evaluation, management, and other aspects, big data + AI can make education more diverse in terms of form and content.

Teaching

"We have to admit that we know too little about students". This is a famous saying of Carnegie Mellon University's School of Education, which is also a common issue in education. For now, production-line education is still common from elementary school to university. Generations of students use the same set of teaching materials, one teacher in charge of one subject, and all are assessed through the same set of standards. Personalized private education is still far from the masses.

Now, big data + AI can help achieve adaptive education and personalized teaching more easily. In terms of teaching approaches, smart classrooms can provide teachers with more abundant teaching methods, full-time interaction, and learning-defined teaching. Teachers no longer have only one textbook in class, but can easily access many high-quality learning resources in the background and present them to students in various forms.

For example, the application of speech recognition and image recognition in education has greatly improved the teaching experience of teachers and students. For a certain English sentence, the teachers can take a photo on the mobile phone and upload it to the cloud. The system will read the sentence with a suitable tone and intonation based on the massive voice material. It can also be combined with the voice evaluation technology, asking students to read the sentence, and the system will access and repeat the reading and scoring.

Moreover, through the combination of virtual reality, augmented reality, and big data, the educational scene can be restored as much as possible, so that students could love and enjoy learning, leading to a greatly improved learning effect. For example, by introducing virtual reality and augmented reality technologies, Google is creating a teaching application called "real-world teaching", which is quietly changing the way classroom activities are conducted.

In the teaching process, by collecting and analyzing the data generated during students' daily learning and homework, teachers can accurately know the mastery of each student's command of knowledge points, and assign homework for each student to achieve the effect of teaching students in accordance with their learning status.

In addition, robot teaching is also a future trend. Previously, the AI robot, Jill Watson, served as a teaching assistant for a month in a class of more than 300 people at Georgia Tech. It can reply to the email immediately with no robot tones, and no one found out it was actually a robot.

Studying

For students during learning, on one hand, big data technology can be applied to make knowledge maps and formulate study plans according to the relationship between knowledge points. On the other hand, data mining technology can help to further analyze students' personal learning level, and establish a learning plan that matches accordingly. The AI system determines how to provide students with personalized supplementary guidance to help them learn efficiently and avoid over-the-top tactics.

For instance, examinations that used to take 3 hours of training may only take 0.5 hour to fully master the knowledge points. Based on this, the application of big data and artificial intelligence can dynamically evaluate students' learnings, and recommend exercises suitable for each student in a targeted manner, which could save time and achieve better learning results.

The image recognition technology can also be applied to improve learning efficiency. Nowadays, students can learn key points and solve difficult problems by uploading the content of teaching materials or homework questions taken on mobile phones. Online classes, encyclopedia links, PPT, PDF files uploaded by teachers, etc. can provide more possibilities for autonomous learning, where the whole process is collected and processed through machine learning and natural language processing technology.

In addition, the development of online education is in full swing. By providing flexible and diverse course forms such as video teaching, riddles, and games, as well as high-quality and rich course content, learning is no longer limited to a certain time and place, but can be arranged flexibly and effectively.

In terms of programming, more and more children are learning through online education. For example, Codemao relies on AI and data mining systems to provide a graphical programming platform for teenagers aged 6–16 with provided differentiated courses. By using graphical programming languages to create games, software, animations, stories, and other works on the platform, students can simultaneously get trained and improve their logical thinking ability, task disassembly ability, interdisciplinary integration ability, and teamwork ability.

Evaluation

In traditional education, examinations and evaluations occupy much of teachers' time. Today, technologies such as big data, text recognition, speech recognition, and semantic recognition are becoming more and more mature, making large-scale automatic correction and personalized feedback a reality.

Through the application of big data and artificial intelligence, the teachers only need to scan the test papers that need to be reviewed, and immediately they can count and display the number of scanned test papers, the average score, the highest score, and the most concentrated wrong questions and corresponding knowledge points in real time.

If one needs to analyze hundreds of thousands or millions of test papers, one can also quickly check all the texts that are similar to the target through accurate image and text recognition and massive text retrieval technology, and then quickly extract and mark the test papers that may have wrong answers. Based on this, intelligent evaluation could be realized.

Management

If to say most learners only pay attention to the "learning" part, then the school education needs to analyze educational behavior data and do a good job of management in addition to teaching. Through intelligent technology, the school can further collect, record, and analyze teaching and learning and related educational behaviors by fully considering the needs of campus management including the Academic Affairs Office, Student

Office, School Office, School Affairs Office, and other departments, and thus better outline the true form of education and teaching, and effectively promote the informatization of teaching.

Currently, some colleges and universities have established applications for student portrait, student behavior warning, student family economic status analysis, student comprehensive data retrieval, student group analysis, etc., to help better distinguish students' potential in professional study or employment direction, and thereby provide them with personalized management and training programs.

For example, facing the diverse course selection needs, how to arrange courses reasonably has become an urgent problem. In the past without artificial intelligence, it often takes several weeks for teachers to arrange courses, yet the satisfaction of all student's needs still can never be guaranteed. AI algorithms are now used to arrange courses, where students only need to submit their own course choices, and the system can quickly arrange courses in combination with courses, classrooms, and teachers. This has greatly improved the arrangement efficiency and student satisfaction.

This is just the beginning for education, where the transformation of big data and artificial intelligence on education will continue. In the future, we will use big data to realize the personalization of education, and use artificial intelligence to empower education. While multiplying the educational capacity, AI will make full use of high-quality teaching resources so that different teaching methods for different students could be realized.

To achieve this, we must not only look up at the starry sky but also stand on solid ground. As educator Ye Shengtao said, education is agriculture, not industry. While education requires time to develop, children also need time to grow up, just like crops. During this process, we believe big data and artificial intelligence will become important nutrients and auxiliary forces for their growth periods.

Chapter 9

See the Future and Take the Lead

With the advent of the era of big data, it has become a global consensus that "whoever gets the data wins the world". Under the new trend of "data-driven", all countries are competing to create data advantages. In this development process, we need to stand in the future and look at the present. In view of the current situation of big data development in China, we should break through the bottleneck of big data development, fully combine big data with artificial intelligence, vigorously cultivate big data talents, promote the healthy development of the big data industry, and finally take this opportunity to get a leg up in the global big data competition.

The Rise of Big Data in China

Today, smart city projects in different parts of the world have put a lot of emphasis on the role of big data, hoping to improve urban public transportation and infrastructure, and achieve a "smart city".

In April 2017, China's Mobike launched "Rubik's Cube", a big data artificial intelligence platform through which millions of bicycles are operated and maintained using big data methods. Mobike's big data application uses the "Rubik's Cube" to monitor various data including vehicle data, cycling distribution data, intelligently recommended parking spot data, and urban cycling demand data. Based on TensorFlow, the second-generation artificial intelligence learning system developed by Google, Mobike can predict the supply and demand of bicycles and schedule vehicles easily. Based on intelligent analysis and reasonable prediction, it can determine the demand for bicycles in a certain location and a certain period of time, so as to more efficiently serve bicycle users in all places.

Overview of the development of big data in various places

At present, many provinces and cities in China regard big data as a strategic emerging industry with key support. Guangdong Province has set up a Big Data Administration to formulate industrial plans and determine the first batch of application demonstration projects. Chongqing has formulated the "Big Data Action Plan", proposing to make the big data industry an important growth point for economic development, and to make itself a big data hub and industrial base with international influence.

At the beginning of 2014, Guizhou Province issued a big data industry development plan and several policies to support it as a key emerging industry. It took big data as the government's "top-in-command" project and established a provincial-level promotion mechanism. In addition, it uses data flow to attract capital, talents, and other elements to build a national-level big data content center, service center, and financial center.

In the same year, Beijing Zhongguancun issued opinions on accelerating the cultivation of big data industrial clusters and promoting industrial transformation and upgrading. The opinion includes the following: improving the policy environment, and gathering innovation resources; building platforms to cultivate big data technology innovation alliances, industry alliances, and other organizations; strengthening regional cooperation and establishing the "Beijing–Tianjin–Hebei Big Data Corridor".

Opportunities brought by the development of big data

China's economic development has entered the new normal, where new-generation information technologies such as big data, cloud computing, and mobile Internet have played an increasingly important role in economic development, as well as "mass entrepreneurship and innovation". The era of big data is running towards us at full speed. We must take this major strategic opportunity to promote national security, social stability, economic development, and people's well-being.

On a global scale, China has the largest population base, the number of people, and data volume. It is estimated that in 2020, the total amount of data in China will reach 8.4 ZB, accounting for 24% of the total global data, making it the world's largest data country and "world data center".

To this end, government departments at all levels should formulate a clear top-level design for big data and cloud computing, starting from key factors such as talents, data sovereignty, key technologies, data research, data innovation capabilities, legal environment support, and industry chains covering the entire industry, study the development trend of big data and evaluate the revolutionary impact of big data on government, economy, and social operation.

Now, big data has occupied half of the Internet, and the development direction of big data has been paid more and more attention to by enterprises. Through the analysis of big data, enterprises can find the most suitable product features that meet the user's needs, and then uses them to guide product design and development. In addition, after the business is launched, it tracks and analyzes users' online pre-orders, usage habits, word-of-mouth evaluation, etc., to provide data support for optimizing business strategies, improve business quality, service level and customer experience, and finally achieve refined network marketing and improve customer satisfaction. Some Internet giants are quite mature in these aspects, and a large proportion of their income is already coming from this. Due to the fact of shortage of talents in related enterprises in the era of big data, universities are intensively launching big data talent training programs.

Big Data Opens Up New Paths for National Governance

Big data is not only a technological revolution and an economic revolution but also a revolution in national governance. As Viktor Mayer-Schönberger, professor of data science at Oxford University, mentioned in his book *The Age of Big Data*: "big data is the source of people gaining new cognition and creating new value, and it is also the way to change markets, organizations, and the relationship between government and citizens".

In April 2018, the "13th Five-Year Plan" proposal pointed out: "using big data technology to improve the timeliness and accuracy of economic operation information". In the era of big data, the Internet has become a new platform for government governance.

Big data is vigorously promoting the modernization of the national governance system and governance capacity. General Secretary Xi Jinping proposed: "the Internet plays a greater role in social management". Premier Li Keqiang also stressed: "We should lock the enforcement power in the 'data iron cage', so that the untrustworthy market behavior has nowhere to hide, and the power operation leaves traces everywhere. This provides the first-hand scientific basis for government decision-making, and realizes 'the cloud knows what people did'".

The development of big data has opened up new ways to modernize government governance. The State Council issued the "Outline of Action for Promoting the Development of Big Data", proposing to establish a management mechanism of "using data to speak, using data to make decisions, using data to manage, and using data to innovate", and gradually realize the modernization of government governance capabilities. The application of big data makes the data on which government governance is based more comprehensive. Collaborative sharing of information between different departments and institutions can effectively improve work efficiency and save governance costs. Government work can better serve the people and can greatly alleviate the problem of people's difficulty in handling affairs.

The Direction of Breaking Through the Bottleneck of Big Data Development

Big data is being widely integrated into all walks of life and has become a new economic growth point. However, in the process of rapid development of big data, there are also problems such as insufficient technological innovation, lack of big data talents, and prominent data security. It is urgent to integrate forces from all walks of life to make breakthroughs.

Emphasize security protection and ensure the big data security

Data are everywhere, and meanwhile, data leakage happens everywhere. According to the "Research Report on Big Data Security Risks and Countermeasures" issued by the Internet Research Center of Shanghai Academy of Social Sciences, since 2013, data security incidents in enterprises or social organizations have frequently leaked more than 100 million records, and the reasons including external attacks and internal leaks, technical and management defects, which have greatly threatened the development of enterprises and personal information safety. Therefore, security mechanisms should be established as long as there exists data.

In terms of information security protection, the government needs to take the lead, and all sectors of society must respond in concert to implement information security. First of all, at the technical level, it is necessary to improve the security protection technology, strengthen research in network security and personal privacy protection, and enhance the ability to defend against information security incident risks. At the same time, government agencies, industry organizations, and large enterprises can set up data governance committees, big data administrations, and other specialized data governance agencies, introduce third-party information security audits, and conduct overall management of data governance.

In addition, in order to clarify the boundaries of data security, we propose to draft and promulgate the Data Law, the Information Protection Law, the Data Open Law, etc., to further improve relevant laws, regulations, and policies, so that data protection can be effectively governed by laws. One should note that in the process of building China's data security system, it is necessary to find a reasonable balance between data security protection and information openness and sharing. It is also necessary to advocate the open sharing of data, so that the production materials in the new era of data can be circulated, and the public can share the social progress and benefits brought by big data technology.

Research and develop key technologies to develop the big data industry

Looking at the global big data competition situation, the big data competition of various countries depends to a large extent on technical strength and innovation ability. Countries with key technologies in the core field of big data are often able to win the big data competition and get a dominant position. Therefore, in the process of vigorously developing the big data industry in China, it is mandatory to strengthen the research and development of key technologies in various links such as real-time data collection, massive information storage and processing, and visual display.

In the processing of big data, big data collection, big data preprocessing, big data storage and processing, big data analysis and mining and other links are all facing the issue of key technologies development, for example, database collection, network data collection, and file collection in the big data collection process; data cleaning, data integration, data transformation, and data reduction in the big data preprocessing process; technology expansion and encapsulation, visualization analysis, data mining algorithm, predictive analysis, semantic engine, and data quality management in big data analysis and mining.

For the above big data technologies, it is urgent for the government to give strong financial and policy support to break through the R&D innovation and application of core technologies, improve the ability of independent R&D and innovation, and build a big data industry chain with core technology autonomy. Especially for core technologies in some key areas, we suggest setting up special financial funds at the national level to break through the bottleneck restricting development.

Innovative training mode to cultivate big data talents

As Didi Chuxing CEO Cheng Wei said, "the bottleneck of big data development is the talents". On March 28, 2017, according to a report by People's Daily, in the next 3–5 years, China needs 1.8 million data talents, yet currently, there are only about 300,000 people, and the talent gap will reach 1.5 million. At the same time, there is an unavoidable fault phenomenon in the big data talent market: the shortage of subject leaders and top talents, while basic talents are also in short supply.

To develop the big data industry, the first step is to train a multi-level, high-quality big data talent team. The government, universities, enterprises, and society need to work together. On one hand, we should focus on training big data application talents that meet the needs of society, not only supporting and assisting undergraduate colleges and universities to open big data-related majors but also for the construction of related majors in colleges and universities, and provide active financial support to train professional skilled talents. On the other hand, we need to actively build a compound big data talent team. On the basis of focusing on training application-oriented talents, we will train talents in the core technology research and development of big data, actively introduce high-end talents, and train a group of compound talents who not only understand data collection, data algorithm, data mining analysis and other professional technologies, but also expert in forecast analysis, market application, management, and command.

Big Data Will Emerge with AI

Big data + AI is the future development trend. In the process of integration of the two, big data will provide "nutrition" for AI, and the developed AI will in turn boost the deep application of big data technology.

Big data provides "nutrition" for AI

In 2016, AlphaGo made a stunning appearance. It is not difficult to find that its success is inseparable from the "feeding" of a large amount of data. AlphaGo has achieved growth and transformation through deep learning in the process of chess and self-play based on massive chess records.

In areas where AI technologies are more and more widely used, such as autonomous driving, graphic image recognition, speech recognition and interaction, and medical diagnosis, they all require massive data to provide a research basis. Based on a large number of pictures, voice, video, and other resources, and with the support of AI technology, the machine has gained the ability to achieve rapid learning and improvement in a short period of time, and can think, make decisions, and act at last.

AI boosts deep application of big data

With continuous breakthroughs in applications such as data analysis models and algorithms, data analysis capabilities have been further improved, where AI can also make more complex decisions. Based on this, the increasingly mature AI technology in turn further promotes the implementation of the deep application of big data.

At present, in this regard, many well-known companies have come to the forefront of the industry. For example, JD.com, in the process of building a smart supply chain management platform, uses core technologies such as machine learning and operational research optimization to establish a data-driven intelligent analysis system to achieve sales forecasting, intelligent replenishment, slow-selling processing, inventory optimization, and other applications.

Opportunities and challenges for humans under the AI wave

In recent years, the rapid development of big data and AI technology has brought great convenience to our daily life. Through AI, Google can scan and identify retinopathy, predict the probability of the patient's onset in advance, and take a diagnostic response as soon as possible. For another example, Google Map can use AR camera for real-time navigation, and easily identify the direction in the street view recognized by AI.

AI is a double-edged sword. While bringing many conveniences and opportunities to people's lives, it also brings many challenges to mankind. As Hawking believes: "artificial intelligence may completely replace humans and eventually evolve into a new form of life that surpasses humans". Yuval Harari, the author of *A Brief History of Humanity*, also said bluntly: "we will be the last generation of Homo sapiens. In another one or two hundred years, the world will be ruled by a completely different entity".

In 2013, Ray Kurzweil, whom Bill Gates believes is the most accurate predictor of AI, made a prediction that the AI singularity is approaching. By 2045, the computer will surpass the human brain, and AI will surpass the human level in an all-round way.

Elon Musk has also published the famous "AI threat theory". He warned that artificial intelligence might cause a third world war. Soon after, he called artificial intelligence the greatest threat to human civilization.

Of course, this is not to exaggerate the AI threat theory but to remind people to think about how to better deal with the changes brought by artificial intelligence. It may be difficult for us to only hope to immigrate to Mars like Musk, but in order to prevent humans from being overwhelmed when super AI comes, it is particularly necessary to strengthen AI security protection research.

The Internet of Everything Enters the Age of Pandora

Avatar depicts a Pandora planet, which is an ecological network where everything is interconnected. All creatures on the planet are connected into an organic whole through a neural network that is more complex than the human brain, thus forming a tight and harmonious ecosystem. In this system, the individual stores all the information of his life in the surrounding plants, where the thoughts of all the plants converge to form an ecological network. Even if a plant dies, its memory and experience will not die out but will be inherited by the entire ecological network. If new life is born, this ecosystem will also give it the wisdom and experience of the entire network, and then make it develop on this basis.

Pandora has the characteristics of the Internet of Everything, yet the core of the Internet of Everything is still the Internet. Currently, the Internet on earth is evolving into an organic whole, where the interweaving of the Internet with 4G, 5G, and the Internet of Things marks that the process of interconnecting everything is accelerating.

What is the Internet of Everything?

The Internet of Everything is the combination of people, processes, data, and things to make network connections more relevant and valuable. The Internet of Everything transforms information into actions, creates new functions for enterprises, individuals, and countries, and brings richer manifestations, infinite possibilities, and unprecedented economic development opportunities.

Today, the Internet of Things has entered the era of "Internet of Everything". All things will gain context awareness, and the processing and sensing capabilities will also be greatly enhanced. Combining people and information to the Internet will create a network of billions or even trillions of connections that brings unprecedented opportunities.

Technology pioneer and founder of 3Com, Robert Metcalfe, coined the famous "Metcalfe's Law": the power of the Internet is growing exponentially as more things, people, data, and the Internet are connected. He believes that the value of the network is proportional to the square of the number of connected users, and the power of the network is greater than the sum of its parts, making the Internet of Everything produce incredible results.

As we enter the era of the Internet of Everything, the public paid more attention to Metcalfe's Law. Everyone wants to know whether it still holds true today, that is, whether the value of the Internet still grows exponentially as the number of users increases.

In fact, the Internet has created unprecedented business opportunities. The silent objects are given voices to awaken almost everything people can think of. The Internet of Everything will give us a clearer picture and a broader perspective, and allow us to make more accurate decisions based on reality.

The development trend of the Internet of Everything

We wake up every day with the convenience of accessing information. Self-driving cars, smart home appliances, robots, etc. have entered the homes of ordinary people from science fiction movies. Artificial intelligence, big data, cloud computing, and the Internet of Things are changing the world at an unimaginable speed, driving consumption upgrades and industrial transformation. Innovation is becoming the tipping point of global industrial transformation.

The important data is becoming the basic resource of new business. We need to seriously think about the question of how to use data to promote the changes of the new economy, the integration of industries, and the reform and upgrading of the supply-side structure. The most essential feature of the Internet is to stay connected to users forever, and then provide services to users. The Internet of Things makes the traditional security concept change from the original network security, computer security, and mobile phone security to life security and user's personal security. For example, when the kid puts on the children's smartwatch, the parents can know their location at any time, and when the child encounters a dangerous situation, the parents can also know it at the very first time.

In fact, the Internet of Everything model has been formed for a long time. However, only after broadband is popularized to a certain extent, and grid computing, management technology, virtualization, fault tolerance technology, and SOA have a certain degree of maturity and integrability, together with the promotion of some large companies, the Internet of Everything would then be able to shine like a new star. Numerous new methods, application of new technologies, and economies of scale, as well as shared use by the public can greatly improve resource utilization, making the Internet of Everything an epoch-making technology.

The Internet of Things uses an astonishing number of sensors to collect massive data, transmit it through 4G, 5G, broadband Internet, etc., and collect it into cloud computing facilities for processing and analysis so that it can be processed and analyzed more quickly, accurately, efficiently, and intelligently manage and control the physical world. Based on this, human beings can manage production and life more finely, and achieve a state of "smartness", thereby greatly improving the productivity and life quality.

To sum up, the future is an era driven by big data, deep learning, and artificial intelligence. The Internet of Everything will change existing business models. Through the Internet of Vehicles, unmanned driving, public safety, medical treatment, and other fields, the use of Internet of Things and "Internet +" can make public utilities more precise. The Internet of Things will turn the small environment and hundreds of millions of objects into the perception and control objects of cloud computing, while 4G, 5G, etc. turn everyone into an application node and intelligent node in cloud computing. The highly integrated cloud computing and grid computing complex is like the Virgin of Pandora, which will unite the power of this planet with its superpower and wisdom.

Promote the Innovation and Development of the Big Data Technology Industry

In recent years, the big data industry has maintained a rapid and healthy development trend, and has continuously penetrated other traditional industries, resulting in a large number of new services, new formats, new products, and new models. With the rapid development of the new generation of the information technology industry and the deepening of information networking in various economic and social fields, the strong domestic application demand and huge market space will provide a stronger driving force for the innovation and development of the big data industry.

However, the big data industry is also faced with several challenges including the need to improve the level of data openness and sharing, the weak leading role of technological innovation in industrial development, the urgent need to improve digital transformation, the urgent need to build industrial statistics and standard systems, and the new problems of data security and data sovereignty. Therefore, we need to take corresponding measures to promote the innovation and development of the big data technology industry. Specifically, we can start from the following aspects.

Strengthen the construction of industrial ecological system

We suggest the following: (1) supporting the development of industrial big data solutions, using big data training to develop new forms of manufacturing in industry, and carrying out pilot projects for innovative applications of industrial big data; (2) promoting the integration of cloud computing, big data, industrial Internet, personalized customization, 3D printing, etc., and promoting the transformation of manufacturing mode and industrial transformation and upgrading; and (3) focusing on accelerating the deep integration of new generation information technology and industry, and actively cultivating new products, new formats, new technologies, and new models.

Support the development of local big data industry

The big data industry in Beijing, Shanghai, Guangzhou, Xi'an, Guiyang, and other regions is developing rapidly, where the first trial and active exploration have achieved initial results. For example, the approval and support of the establishment of the Guiyang-Gui'an big data industry development cluster area has achieved remarkable results in the introduction of industrial support policies, data-sharing transactions, laws, regulations, etc. In the future, the Ministry of Industry and Information Technology will further mobilize and support all localities, industries, and departments to carry out exploration and practice of big data technology, industry, application, policy, and other aspects. Moreover, they will rely on relevant project funds to carry out application demonstrations in key areas and industries, summarize experience, deepen reform, and accelerate promotion.

Explore and strengthen industry management

We should combine with the strategies of "building an Internet power" and "broadband China" to guide the scientific layout of data centers, accelerate the promotion of broadband popularization, and improve the market management level of Internet data center business. In addition, we should increase the management of privacy information protection, network security, and cross-border data flow. Further, we should promote and cooperate with relevant departments to organize legislative research work on data sharing, openness, security, and transactions. Lastly, we will need to address the system and market uncertainty factors in the development of the big data industry, to create a good regulatory and market environment for industry and application.

Strengthen core technology innovation

In addition to strengthening big data core technology research, and laying out major national big data science and technology innovation projects, it is also necessary to accelerate the transformation of scientific research achievements, and develop interdisciplinary and cross-field technology research driven by application needs, gathering resources from multiple parties to jointly accelerate the industrialization process of cutting-edge big data technologies. At the same time, we should build an innovation grid that supports digital transformation, and coordinate the construction of national big data comprehensive pilot areas, industrial agglomeration areas, and new industrialization demonstration bases, support innovation and entrepreneurship in the field of big data application, and encourage and support small and medium-sized and start-up enterprises to strengthen the development of big data application technology.

Promote the development of the real economy

In order to conduct the research and development and enterprization of industrial big data technologies, products, platforms, and solutions, and build a number of national, industry-level, and enterprise-level industrial Internet platforms, it is necessary to promote the application of big data in the fields of computing science, resource exploration, satellite applications, modern agriculture, and major equipment manufacturing, as well as to promote the integration of big data with business, finance, medical care, culture, education, and other fields.

By 2020, it is expected that the global data usage will reach about 40 billion terabytes, covering various fields such as economy, society, military, and politics, and becoming a new important driving force for the world's development. Data are a very precious asset of the 21st century. Big data is changing the comprehensive national strength of various countries, reshaping the future international strategic pattern, and redefining the space for games between major powers. In the era of big data, countries in the world are increasingly dependent on data, and the focal point of national competition has shifted from the competition for capital, land, population, and resources to the competition for big data.

China will make better use of the Internet, big data, and cloud computing to provide a platform for mass entrepreneurship, and jointly promote data openness, data security, industrial prosperity, technological innovation, and integrated development.

In fact, how to develop big data has become an important topic for the country, society, and industry. In the future, as China vigorously promotes the integration of the big data industry and public services, it will effectively promote the upgrading of industrial quality and efficiency, and improve the level of government governance and public service capabilities. Big data belongs to everyone, every company, and especially the whole country. With the implementation of the national big data strategy, smart life, smart enterprise, smart city, smart government, and smart country based on big data will eventually be realized.

Big Data Talent Training

The "Big Data Industry Development Plan (2016–2020)" released in December 2016 pointed out that "there is a shortage of talents in big data basic research, product development, and business applications, which is far from demanded".

With the application of big data technologies such as data collection, data storage, data mining, and data analysis in more and more industries, the shortage of big data talents has become an increasingly prominent issue. McKinsey predicts that the number of fresh graduates in data science will increase by 7% each year. However, the demand for professional data scientists in high-quality projects alone will increase by 12% each year. The talent will be in badly short supply.

Taking Guizhou University as an example, the employment rate of its first graduate students majoring in big data reached 100%. The urgent demand for talents has directly stimulated the major of big data. The Ministry of Education (MOE) officially established the undergraduate major of "Data Science and Big Data Technology" (professional code: 080910T) and the junior major of "Big Data Technology and Application" (professional code: 610215).

According to the notice of the 2017 filing and approval results of undergraduate majors in colleges and universities announced by MOE, the number of colleges and universities that applied for and are approved to have the data science and big data technology majors has increased to 278. On January 18, 2018, the MOE announced the professional filing and approval results of "Big Data Technology and Application". A total of 270 vocational colleges have applied for the "Big Data Technology and Application" major, of which a total of 208 vocational colleges have been approved. With the development of big data, the number of colleges and universities applying for and being approved for this major will continue to increase in the next few years.

However, for now, in terms of big data talent training and big data course construction, most colleges and universities are still in their infancy, and there are still many issues to be explored. First of all, big data is a new thing, there are very few teachers who understand big data, so good teachers are in demand. Second, a perfect big data talent training and curriculum system has not been formed, so a good mechanism is in demand. Third, big data experiments need to provide cluster computers for each student, so machines are in demand. Finally, colleges do not have massive data, so the "raw materials" to carry out big data teaching and research work are in demand.

Bibliography

Ali Research Institute. PAN Yonghua of Ali Research Institute: 5 directions for big data to promote the transformation and upgrading of the manufacturing industry. *Hangzhou Keji*, 2016(4): 43–46.

Analysis on the influence and countermeasures of big data in China e-government. http://news.hexun.com/2017-05-11/189150871.html.

Armbrvst M, Fox A, Griffith R, *et al*. Above the clouds: A Berkeley view of cloud computing. *UC Berkeley, RAD Laboratory*, 2009.

Brewer E. Towards robust distributed systems. Keynote at the *ACM Symposium on Principles of Distributed Computing (PODC)*, 2000.

Che Hongying. Research on the construction and development path of smart government affairs in the era of big data. *Accounting Learning*, 2017(15): 203.

Chen Jiayou. Research on the experience and enlightenment of foreign big data development strategy, 2017(26): 12–13.

Chen Shimin. Big data analysis and data velocity. *Journal of Computer Research and Development*, 2015, 52(2): 333–342.

Chen Wan. Thinking and exploration of public security work under the background of big data era. *Legal System and Society*, 2015(15): 202–203.

Chen Wei. Interpretation of "notice of the state council on issuing the action outline for promoting the development of big data". http://www.ithowwhy.com. cn/auto/db/detail.aspx?db=999021&rid=37828&md=1&pd=12&msd=1&ps d=12&mdd=7&pdd=213&count=10. 2015-09-10.

Chen Xiaoguang. Survival, evolution, and transformation: The only way for urban radio and television. *China Radio & TV Academic Journal*, 2018(4): 42–45.

Chen Yan, Zhang Jinsong. *Big Data Technology and Its Applications*, Dalian Maritime University Press, 2015.

Chen Yuzhao. Accurate efforts to attract big and strong. *Guizhou Daily*, 2016-05-25.

Dai Chunchen, Wang Pengjun. In the future, every teacher will need an AI assistant. *21st Century Business Herald*, 2017-08-22.

Development prospects of data analysts: Industry chapters of big data application scenarios. https://www.cda.cn/view/118467.html.

Duan Yunfeng, QIN Xiaofei. Internet thinking of big data. Publishing House of Electronic Industry, 2015.

Duo Shujin, GUO Mei. Problems and countermeasures of smart government construction in China. *Journal of Baoding University*, 2015, 28(5): 38–43.

Evelyn Zhang. Citymapper's first commercial bus route to ease urban traffic and Uber's snatch for big data. http://www.p5w.net/news/cjxw/201707/t20170724_1890010.html.

Feng Boqin, GU Gang, XIA Qin, *et al. University Computer Foundation (Windows XP + Office 2003)*. Beijing: Posts & Telecom Press. 2009.

Fu Jian, Pu Junyi. Computer industry: Interpret big data, tell a truest story. Everbright Securities. https://wenku.baidu.com/view/ee6dab09ba1aa8114 431d9bb.html.

Galaxy Statistics Studio, Harbin University of Commerce. Network statistical graphics. https://www.cnblogs.com/cloudtj/p/6097451.html#A1.

Global network and cooperative banks deeply cultivate small and micro industries in the business circle. http://dengym_qqwcom.blogchina.com/2118502.html.

Guo Yuhui. How to deal with big data? What are the processing modes of big data? http://plus.tencent.com/detailnews/892.

He Ping. The thinking change of leaders in the era of big data. *Journal of Dalian Official*, 2016(4).

Hey T, Tansley S, Tolle K. The Fourth Paradigm: Data-Intensive Scientific Discovery. *Proceedings of the IEEE*, 2011(8): 1334–1337.

Hey T, Tansley S, Tolle K. The Fourth Paradigm: Data-Intensive Scientific Discovery. Redmond: Microsoft Research, 2009.

Hu Qianwen. Make government services "zero distance". *Yuxi Daily*, 2017-08-28.

Huang Xin. How big data affects traditional industries. *Economic Daily*, 2017-07-07.

Huang Xin. China's big data world ranking. *Economic Daily*, 2017-07-09.

Introduction to the application scope and purpose of various Excel charts. https://yq.aliyun.com/wenji/166289.

Introduction to Yan'an 12345 smart government service and social situation polling platform. http://12345.yanan.gov.cn/about.php?cid=7.

Ju An. What happens when cities have brains. *Innovation Time*, 2018(01): 10–11.

King of data. A picture to understand how big data has reached the "high level" of the national development strategy. http://www.cbdio.com/BigData/2016-01/29/content_4566055.htm.

Large-scale graph computing research. http://www.360doc.com/content/18/0715/14/57515853_770556529.shtml. 2018-07-15.

Let data help the city to think and make decisions, Hangzhou uses the "city brain" to manage traffic problems. *Auto & Safety*, 2018(02): 77–79.

Li Bo, DONG Liang. The mode and development of Internet finance. *China Finance*, 2013(10): 19–21.

Li Jieyi, LIU Yuqin. Big data technology supports innovation of targeted poverty alleviation models. *Chinese Social Science Today*, 2017-09-27.

Li Junjie. FB, CLOUD, GDPR and More Privacy — Protection and data security in the era of big data. https://mp.weixin.qq.com/s/hQywQWLtW8pncv_VAQcYsw. 2018-06-10.

Liang Liyan. Application research of association rule mining apriori algorithm in digital archives system. *Modern Computer*, 2011(07): 7–10.

Lipcon T. Design patterns for distributed nonrelational databases. Cloudera, 2009.

Liu Chao. Finally, someone understands cloud computing, big data, and artificial intelligence. https://sq.sf.163.com/blog/article/217814081753378816. 2018-11-05.

Liu Peng, Yu Quan, Yang Zhenyu, *et al. Cloud Computing for Big Data Processing*. Beijing: Posts & Telecom Press. 2015.

Liu Peng. Cloud computing opens the age of Pandora. https://www.tvoao.com/a/33299.aspx.2011-08-16.

Liu Peng. *Cloud Computing*, 3rd Edition. Beijing: Publishing House of Electronic Industry, 2015.

Liu Peng. *Big Database*. Beijing: Publishing House of Electronics Industry, 2017.

Liu Peng. *Big Data Visualization*. Beijing: Publishing House of Electronics Industry, 2018.

Liu Peng. *Big Data*. Beijing: Publishing House of Electronic Industry, 2018.

Liu Runda, SUN Jiulin, LIAO Shunbao. A preliminary study on data authorization in scientific data sharing. *Journal of Intelligence*, 2010, 29(12): 15–18.

Lu Jinhong. Introduction to classification algorithms. https://blog.csdn.net/jediael_lu/article/details/44152293. 2015-06-16.

Machine learning — Summary of classification algorithms. https://blog.csdn.net/lsjseu/article/details/12350709. 2013-10-06.

Meng Xiaofeng, CI Xiang. Big data management: Concepts, techniques and challenges. *Journal of Computer Research and Development*, 2013, 50(01): 146–169.

Meng Xiaofeng. *Introduction to Big Data Management*. Beijing: China Machine Press, 2017.

Ministry of Industry and Information Technology. Big data industry development plan (2016-2020). http://www.miit.gov.cn/n1146285/n1146352/n3054355/n3057656/n3057660/c5465614/content.html. 2016-12-18.

Mobike launches "Cube" platform, big data intelligent operation and maintenance is the key. http://news.cnfol.com/it/20170413/24585796.shtml. 2017-04-14.

Nie Guangli, JI Xiaotian. Research on Internet credit model and suggestions for commercial banks. *Rural Finance Research*, 2015(2): 18–23.

Niu Wenjia, LIU Jiqiang, SHI Chuan, *et al. User Network Behavior Profile: User Network Behavior Profile Analysis and Content Recommendation Application in Big Data*. Beijing: Publishing House of Electronics Industry, 2016.

Outline of the 13th five-year plan for the national economic and social development of the People's Republic of China. https://wenku.baidu.com/view/1286562889eb172dec63b799.html. 2016-03-17.

Peng Wenmei. Analysis on the co-construction and sharing of university library resources in the era of big data. *Journal of Zhaoqing University*, 2014(01): 86–90.

Qiu Wenbin. Big data creates "smart government affairs". http://www.jlth.gov.cn/nd.jsp?id=76. 2018-10-24.

Ruanyifeng. Data visualization: Basic charts. http://www.ruanyifeng.com/blog/2014/11/basic-charts.html.

Shan Yong. Using data governance to innovate the social security prevention and control system. *Studies on Socialism with Chinese Characteristics*, 2015, 1(4): 97–101.

Shang Yang. *The Book of Lord Shang*. ZHANG Jie, translated. Beijing: Beijing United Publishing Co., Ltd., 2017.

Shenzhen Baoan's "smart government affairs" helps the construction of modern service-oriented government. *Computer & Network*, 2016, 42(13): 7–9.

Shi Qu. Graph computing engine - Bringing association analysis to the limit. https://www.prnasia.com/story/166062-1.shtml. 2016-12-12.

Summary of classification algorithms. https://blog.csdn.net/chl033/article/details/5204220. 2010-01-17.

Tang Guochun, Luo Ziqiang. Multi-level research in cloud computing architecture. *Railway Computer Application*, 2012, 21(11): 4–7.

Three common data normalization methods. https://blog.csdn.net/bbbeoy/article/details/70185798. 2017-04-15.

Tian Kuang, Du Ninglin. An ensemble clustering algorithm for high-dimensional data. https://cloud.tencent.com/developer/article/1057671. 2018-03-12.

Types, comparisons, and choices of big data processing frameworks. https://blog.csdn.net/huangshulang1234/article/details/78640938. 2017-11-29.

Victor Meier-Schönberger, Kenneth Cookyer. *The Era of Big Data*. SHENG Yangyan, ZHOU Tao, translated. Hangzhou: Zhejiang People's Publishing House, 2013.

Wang Li, Jiang Gang, LI Jianping, China comment: The rise of China's big data industry. http://article.ccw.com.cn/article/view/102601.

Wang Lu. *Big Data Leading Cadre Reader*. Beijing: People's Publishing House, 2015.

Wang Qiang, Lin Licheng, Zhao Chang. Create a new mechanism for high-level talent evaluation and accelerate the transformation of new and old kinetic energy. *Chinese Talents*, 2017(01): 302–308.

Wang Yanying. The role of "river chief system" in river governance. *Journal of Teaching*, 2017(27): 66.

Willtongji, Data science, data technology and data engineering. https://blog.csdn.net/willtongji/article/details/52874536.2016.

Wu Hequan. The era of "big wisdom moving cloud" is coming. http://news.sciencenet.cn/htmlnews/2017/3/371416.shtm?id=371416. 2017-03-22.

Xie Xuefeng. Using big data to promote the construction of smart government. *Science Technology Vision*, 2016(2): 233.

Xiong Chunlin, Li Hui. Countermeasures on rural public crisis prevention and control in big data era. *Journal of Library and Information Science of Agriculture*, 2018(05): 75–79.

Xu Jinye, Xu Lin. ERP builds accounting big data analysis platform: The core of enterprise accounting cloud computing construction. *Friends of Accounting*, 2013(24): 97–100.

Xu Zongben. Make good use of big data requires great wisdom. People's Daily, 2016-03-15.

Yang Hong. Research on "cloudization" of accounting information system. *Financial Accounting*, 2015(11): 5–9.

Yang Wen. Thoughts on Guizhou's relying on big data to promote economic and social development. *Journal of the Party School of Guiyang Committee of CPC*, 2015(5): 8–11.

Yang Xiuping. Research on the application of big data in Internet financial risk control. *Electronics World*, 2014(17): 12–13.

Yao Liya. Public opinion guidance for emergencies in the era of big data. *Journal of News Research*, 2015(15): 20–21.

Ye Yurui. Characteristics of enterprise-level storage in the era of cloud computing. *Journal of Guangzhou City Polytechnic*, 2016, 10(3): 37–47.

Zhan Zhiming, Yin Wenjun. Environmental big data and its application in environmental pollution prevention and management innovation. *Environmental Protection*, 2016, 44(06): 44–48.

Zhang Chaojun, Chen Huizhong, Cai Jinan, *et al*. Discussion on some problems of earthquake early warning engineering. *Journal of Engineering Studies*, 2014, 6(04): 344–370.

Zhang Chengbo. Learning and applying "big data" to improve the efficiency of E-government. *Fujian Quality Technology Supervision*, 2013(6): 27–28.

Zhang Jianlin. *Management Information System*. Hang Zhou: Zhejiang University Press, 2014.

Zhang Jun, Yao Fei. Research on the construction of national innovation system in the era of big data. *Forum on Science and Technology in China*, 2013(12): 5–11.

Zhang Monan. Constructing a national security strategy in the era of big data. *China Outsourcing*, 2015(07): 76–80.

Zhang Ruixin, Liu Hongbo. E-government anti-corruption effectiveness performance and promotion strategy. *Administration and Law*, 2013(10): 1–4.

Zhang Tiankan. In the era of big data, the game between data use and privacy protection. Guangming comments. http://guancha.gmw.cn/2017-03/09/content_23930917.htm.

Zhang Wenbin. Talking about big data and public security informatization construction. *Informatization Construction*, 2016(06): 32.

Zhang Xiaoqiang, Zhang Yingqiang. How to better implement the national big data strategy research. *Globalization*, 2018(01): 30–41.

Zhang Xinhong, Yu Danxia, Liu Libing, *et al.* The basic framework of the information city development strategy. *E-Government*, 2012(Z1): 20–30.

Zhang Zhenzhu, Zhang Li. The evolution of the food safety supply chain under the background of big data era. *Food Research and Development*, 2014, 35(18): 209–212.

Zhao Yinhong. Smart government: A new direction for the development of E-government in the era of big data. *Office Informatization*, 2014(22): 52–55.

Zhong Ying, Zhang Hengshan. The origin, impact and countermeasures of big data. *Modern Communication (Journal of Communication University of China)*, 2013, 35(07): 104–109.

Zhou Yi, Li Yanjuan. *Database Principles and Development and Application*, 2nd Edition. Beijing: Tsinghua University Press, 2013.

Zhou Zhihua. *Machine Learning*. Beijing: Tsinghua University Press, 2016.

Zhu Jie, Luo Hualin. *Big Data Architecture: From Data Acquisition to Deep Learning*. Beijing: Publishing House of Electronics Industry, 2016.

Zhu Xinjuan. *Database Technology and Application Based on VFP and SQL*. Xi'an: Xidian University Press, 2004.

Index